History of the Alps 1500–1900

HISTORY OF
The Alps 1500–1900
environment, development, and society

JON MATHIEU
translated by Matthew Vester

MORGANTOWN 2009

West Virginia University Press, Morgantown 26506

© 2009 by West Virginia University Press

All rights reserved

First English edition published 2009 by West Virginia University Press

Printed in the United States of America

16 15 14 13 12 11 10 09   1 2 3 4 5

ISBN-10 1-933202-34-3

ISBN-13 978-1-933202-34-1 (alk. paper)

Mathieu, Jon.

[Originially published as *Geschichte der Alpen, 1500–1900*. (German edition, 1998]

History of the Alps, 1500–1900 : environment, development, and society / Jon Mathieu;

translated by Matthew Vester. -- 1st ed.

p. cm.

ISBN-13: 978-1-933202-34-1 (pbk. : alk. paper)

ISBN-10: 1-933202-34-3 (pbk. : alk. paper)

1. Alps--History. 2. Alps--Economic conditions. 3. Economic development--Alps--History.

4. Alps--Environmental conditions. 5. Alps--Population. I. Title.

DQ823.5.M3813 2009

949.4'7--dc22

Library of Congress Pre-assigned Control Number: 2009004105

Cover layout by Rebecca Lown / Book Design and Art Direction by Than Saffel

Typesetting by Michael Rabjohns/PageFlow

Table of Contents

Preface I

1. **The Alps: A Historical Space?** 5
   Key questions and the state of the research 8
   The political construction of territory 14
2. **Population** 23
   Data and collection methods 24
   Comparing long-term trends 34
3. **Agriculture and Alpiculture** 47
   The intensity differential in the Alps 49
   Cropping frequency and yields 55
   The intensification of animal husbandry 61
   . . . and of plant cultivation 64
   Technology 70
4. **Cities** 83
   Statistics in the early modern era 85
   Acceleration of growth 92
   The slowing of urban growth 97
   The nineteenth century 104

5.	**Environment and Development**	114
	An intermediate assessment: differentiated growth	115
	Relations between the Alps and surrounding areas	119
	History and ecological models	127

	Illustrations	After p. 134

6.	**Two Agrarian Structures (Nineteenth Century)**	135
	Farming establishments	137
	Public order and property	146
	Inheritance law, collective resources	154
7.	**Territories during the Early Modern Period**	161
	Savoy: the duke, the notables	163
	The Grisons: communes with subjects	171
	Carinthia: Lord, peasant, servant	180
8.	**State Formation and Society**	195
	The European dimension	196
	Politics as a factor of differentiation	205
	Rural societies	209
9.	**History of the Alps from 1500 to 1900**	222
	A summary	222
	Arguments and outlook	225

	Appendices	229
	Bibliography	241
	Index	250

Maps

Map 1.1: Physical relief of the Alps 10

Map 1.2: Political divisions in the Alps, before 1790 and in 1900 18

Map 2.1: Regions whose surface area is 75–100% Alpine 27

Map 3.1: Forms of agriculture in the Alps, ca. 1900, according to
de Martonne 54

Map 4.1: Cities with 5000 or more inhabitants, in the Alps and surrounding
areas, 1500 87

Map 4.2: Cities with 5000 or more inhabitants, in the Alps and surrounding
areas, 1800 88

Map 4.3: The area surrounding Innsbruck, 1928 100

Map 6.1: Medium and large farms in Alpine regions and districts, 1900 140

Map 6.2: Agricultural servants in Alpine regions and districts, 1900 141

Map 6.3: Illegitimacy rates in Alpine regions and districts, 1900 142

Map 7.1: The duchy of Savoy in the late eighteenth century 165

Map 7.2: The Freestate of the Three Leagues before 1797 173

Map 7.3: The duchy of Carinthia in the late eighteenth century 182

Map 8.1: Elements of agrarian structure in the Alps and in surrounding areas,
eighteenth and nineteenth centuries 207

Tables

Table 2.1:	Indications of total Alpine population levels, 1500–1900	26
Table 2.2:	Population in Alpine regions, 1500–1900	28
Table 2.3:	Population growth and density in Alpine regions, 1500–1900	37
Table 2.4:	Population growth and density in the Alps and in surrounding areas, 1500–1900	38
Table 3.1:	Introduction of corn and potatoes in the Alpine regions, 1500–1900	67
Table 4.1:	Cities with 5000 or more inhabitants, in the Alps and at the foot of the Alps, 1500–1800	89
Table 4.2:	Cities and urban population in the Alps according to altitude, 1600–1800	103
Table 4.3:	Cities and urban population in the Alps according to size, 1800–1900	105
Table 4.4:	Cities and urban population in the Alps according to altitude, 1800–1900	107
Table 7.1:	Household and family in Savoy-Piedmont and in the Grisons, 1561–1832	178
Table 7.2:	Household and economic status in Carinthia and in the Grisons, 1750–1798	187
Table A.1:	Percentages of agricultural workers in Alpine regions, 1870 and 1900	231
Table A.2:	Farm sizes in Alpine regions, 1900	232
Table A.3:	Agricultural labor force in Alpine regions, 1900	233
Table A.4:	Illegitimacy rates in Alpine regions, 1870 and 1900	235
Table A.5:	Farm size, agricultural labor force, and illegitimacy rates in Alpine regions of Hapsburg Austria by district, 1870 and 1900	236
Table A.6:	Cities with 5000 or more inhabitants in the Alps, 1870 and 1900	238

Preface

This book examines the history of the Alps from the end of the Middle Ages until 1900 and is guided by two questions: What was the nature of the relationships between demographic growth, economic development, and Alpine environment? And in what manner did political factors affect agrarian structure and society? The first question is essentially economic, and provides the theme of chapters two through five. The second one is sociopolitical, and is taken up in chapters six through eight. In both sections our methodological point of departure is the premise that the history of the Alpine space is best understood if it is examined in the context of the low-lying areas that surround it. In closing, the results and arguments will be summarized in an overview. But before beginning, it would be helpful to explain why historians became interested in a territory that is defined above all in geographic terms: the Alps, a historical space?

"Perhaps the last secret of the impression created by the high Alps is a sense of distance from life," wrote, just about a hundred years ago, Georg Simmel, as he described the feelings generated by the mountains' mass, form, and human absence. The German philosopher and sociologist was known for his observational and interpretive abilities. Like many other intellectuals of the period who loved to travel, he had experienced the mountains personally. Last secrets seem to me to be too secret. But the distance

Preface

mentioned by Simmel and, even more so, the public emphasis on this distance are both reasons for my persistent restless interest in Alpine matters.

This began with the history of a valley during the Old Regime, a project that weighed on me for a long time, as page after page was filled with text. Later there followed a history of agriculture during the early modern period, in a larger Alpine region. Now I have in front of me this book that stretches from the sixteenth through the nineteenth centuries and deals with the entire Alpine arc. This kind of journey through time and space might at first seem linear, if not unimaginative. A closer look, though, shows that it is only partly a story in installments. Many of the themes that I examined in my previous studies do not play any role in this one. Rather, attention is focused here on issues that had been either only briefly mentioned or totally ignored in my earlier work. On the whole, the move from a single Alpine valley to the entire Alpine space, or from microhistory to macrohistory, has narrowed down the problem set. Readers will not find here a broad description of daily life in the Alps between 1500 and 1900, but a study concentrating on specific topics, chosen because of their general significance, and thus also because of their theoretical interest.

The conception of the book made it advisable to focus on economic and sociopolitical processes and to address cultural elements only sporadically. As a historian who is convinced that it is fruitful to alternate approaches, I believe that it would be misleading to represent this decision, here, as a question of principle. On the other hand, for objective reasons, some of the themes that are at the forefront of the literature play a secondary role in this account. These include transalpine traffic, incipient tourism, and the ever-growing exploration and investigation of the mountains. What unites these topics, which are fascinating in themselves, is the fact that during the period of this study they affected a relatively small number of people, and seem more important from the perspective of the lowlands than from the point of view of the Alpine population. This having been said, I do not wish to insist on a pure and simple internal perspective, since such could distort past reality as much as the traditional external perspective has.

A whole series of persons has inspired me and supported my work, and I thank them for this: Dionigi Albera, Werner Bätzing, Heinrich Berger, Jean-François Bergier, Ester Boserup, Jean-Pierre Brun, Christoph Brunner, Pierre Dubuis, Hans-Rudolf Egli, René Favier, Paul Guichonnet, Peter Hersche, André Holenstein, Kurt Klein, Franz Mathis, Jakob Messerli, Darja Mihelic, Claude Motte, Arnold Niederer, Anne Radeff, Guglielmo Scaramellini, David J. Siddle, Ferruccio Vandramini, Pier Paolo Viazzo and, for this English edition, Matt Vester. I am also grateful to the students who have shared their ideas, orally or in written form, about the topics examined here. None of them, naturally, is responsible for the affirmations or omissions of the present text.

Another word of thanks goes to the academic and state institutions of the countries in the Alpine arc for their various forms of assistance. They have all been important; I must limit myself here to pointing out the inestimable town library of Burgdorf, directed by Ziga Kump. The Swiss National Science Foundation underwrote a large portion of the costs. The balance was probably paid by the passion for intellectual activity, or for that distance from life, as experienced in study offices and libraries. This also included Felicitas, Cla, and Luisa, whose patience with the mountains will be thanked in a different way, in accordance with their wishes.

1 The Alps: A Historical Space?

The winter of 1988–89 witnessed two events whose importance for the history analyzed in this book was greater than one might imagine. In November 1988 the presidents of the three Alpine working communities, ALPS ADRIATIC, ARGE ALP, and COTRAO, met in Lugano, Switzerland for a conference that was intended to be the first in a series of meetings between regional administrative leaders. In an unrelated development, the German minister of the environment announced in January 1989 that he planned to invite his counterparts from all of the countries with Alpine regions to a conference at Berchtesgaden. The conference was held later that year. Then, on 7 November 1991, these ministers signed a framework agreement for an Alpine Convention whose details remained to be worked out. By this agreement, the countries involved (Austria, France, Germany, Italy, Yugoslavia [later Slovenia], Liechtenstein, Switzerland, and eventually Monaco), along with the European Community, committed to a specific policy of environmental protection and development for their Alpine regions. The detailed definition of the territory in which they pledged to follow this policy was itself included in the Convention agreement. Despite (or even because of) its definition by state officials, this territory was much more a function of geographic data than were the territories of the aforementioned Alpine working communities from across the Alpine arc. The territories represented by the working communities included both real Alpine zones and large amounts of surrounding areas—examples included the Yugoslav republic of Croatia and the entire Rhône-Alps region in France. However one wanted to define it, in the years around 1990 the Alpine arc became, for the first time in its history, a space in which the first elements of a common political and administrative structure were being developed.[1]

The preliminary conditions for these events are to be found in representations of the Alps and related discourses whose roots reach far back in time. But at the institutional level, these new structures were a product of the postwar years, after 1970 in particular. The organization that provided the chief impetus for the Alpine Convention, CIPRA (The International Commission for the Protection of the Alps), was founded in 1952 as an offshoot of the International Union for the Protection of Nature (which had in turn been created in 1948 by UNESCO). During the 1950s and 1960s CIPRA became involved in issues relating to the protection of nature in transnational contexts, but it remained a small organization that was dependent on initiatives taken by single individuals, and its activities eventually stopped. It was revived in 1974 as a result of an international symposium in Trent, Italy, dedicated to "The Future of the Alps." This meeting functioned both as a vehicle for activating a renewed, broader commitment, and as an expression of this commitment. Two years earlier, the Working Community of the Alpine Regions (ARGE ALP) had been created in Innsbruck, bringing together the federal districts of the Tyrol, Vorarlberg, and Salzburg in Austria; the state of Bavaria in Germany; the canton of Grisons in Switzerland; the autonomous province of Bolzano and the region of Lombardy in Italy. Later the administrations of other regions were added. Regular exchanges on all manner of public affairs were to be carried out via meetings of heads of state and senior officials. To this end the president of the region of Lombardy organized a sizable conference on "The Alps and Europe" in Milan in 1973. And, based on the model of the Working Community, two other communities (ALPS ADRIATIC and COTRAO – the Western-Alps Working Community) were established in 1978 and 1982, respectively.[2]

The historical context of these activities is the process of European integration and the consequent reconfiguration of nation-state centers and border regions. The movement toward regionalism expressed itself prominently in the Alpine zone, which is crisscrossed by numerous national borders. But the regionalist movement in the Alps also remained somewhat abstract and administrative. What real points of contact could there be between citizens of the Austrian *Land* of Carinthia and those of the distant Maritime Alps in France? How could a common sense of belonging be sparked at meetings of the presidents of working communities that remained largely unknown? It is true that from the 1970s onward, a number of spontaneous initiatives testified to a heightened sense of Alpine self-consciousness. For example, in 1987 widespread opposition to the construction of a new artificial lake prompted people in various places across the Alps to invent an annual tradition of lighting bonfires on a particular weekend in August, in order to demonstrate the "vital cultural autonomy of the Alps." And in the summer of 1992 a group of activists hiked from Vienna to Nice, highlighting problems shared by different Alpine regions and creating networks of like-minded people stretching from one end of the Alps to the other. Still, compared to other kinds of transnational political movements that were linked to specific institutional contexts and thus varied significantly from one region to another, these initiatives had a limited impact.[3]

From the beginning, scholarly research played a significant role in this new orientation, both independently of the political agenda, and in support of it. The disciplines represented by such research runs the gamut from the natural sciences to the humanities, although geography has often played the leading role. The discipline of geography had a long-established interest in defining the Alpine region in physical terms, boasting a proud monographic tradition on this topic. Alongside the politicization of Alpine space there appeared a new generation of general overviews, texts whose titles often clarified their authors' public positions, such as Paul and Germaine Veyret's *Au coeur de l'Europe – les Alpes* (*In the Heart of Europe – The Alps*, 1967), and Werner Bätzing's *Die Alpen. Naturbearbeitung und Umweltzerstörung. Eine geographisch-ökologische Untersuchung* (*The Alps. Transforming Nature and Destroying the Environment. A Geographic-Ecological Inquiry*, 1984). These kinds of analyses of current problems and future possibilities also resulted in a specific set of ideas about the past. In this sense, Alpine regionalism is not unique. The 1974 Trent symposium did not include any real involvement by a single professional historian. Nonetheless, it approved an action plan whose introduction read: "As part of Europe's heritage, the Alps constitute a natural, historical, cultural, and social unity of vital importance. In all of their parts, the Alps have played a decisive role, separating the great currents of civilization while also transforming and integrating them. Notwithstanding interactions and unions—among Alpine peoples and political systems—that were sometimes difficult, an autonomous Alpine culture has developed, and although the Alps have never been politically unified, the lifestyles and activities of its peoples demonstrate the peculiar characteristics of a surprising affinity."[4]

When faced in such unmistakable fashion with the need for a specific historical image, history as a discipline can respond in various ways. One response is to emphasize its own autonomy and to deal with different issues, those stemming from traditions and discussions within the discipline. This is an important strategy: it makes more room available for creative efforts, thereby also increasing the ability of research to respond to the possible future demands that may be placed on it. On the other hand, our discipline can also take under consideration questions derived from without, from the social sphere. It can grapple with such questions while also maintaining a critical distance from the immediate context that gave rise to them—seeing them as opportunities for fresh scholarly discovery—through the generation of new historical narratives.[5] This is my approach in this book.

Key questions and the state of the research

This work examines the history of the Alps between the sixteenth and the nineteenth centuries through the framework of two central perspectives, an economic one and a sociopolitical one.

(1) How did demographic expansion, economic development, and the Alpine environment affect each other? The foundation for the investigation of this theme is provided by an assessment of the population trends in different parts of the mountain range (chapter 2). Against this demographic backdrop, the significant agricultural sector is examined, and then the urban one—which, although limited, was especially important in certain ways (chapters 3 and 4). Here also the impact of environmental factors on economic development is taken into account, and chapter 5 summarizes perspectives on this topic. A key part of my argument is that population growth not only strained scarce resources (a point that is often made in studies of mountain areas); it also functioned as an engine of agricultural intensification and urbanization. That is, it also increased the potential for economic expansion. The main question is: how and to what degree were these phenomena possible under Alpine conditions? This question cannot be answered according to a methodology whose point of departure limits itself to geographic variables. A historical account of actual economic development must be central to this analysis.

(2) In what ways did political factors influence the structures of rural societies? Unlike the first perspective, this one draws attention to differences existing within

the Alpine world. Its starting point is an observation about agricultural structures: a feudal/large peasant farm system was predominant in eastern regions of the Alps, and differed sharply from the village-based/small-farmer model prevalent in other Alpine areas. This fact is perhaps more significant than any other internal difference that can be identified in the Alps. The composite qualifiers used to describe these phenomena themselves indicate that if we are to understand the historical conditions that gave rise to agrarian and land-holding structures, we must be more attentive to politics than to economics. In order to provide an Alpine-wide statistical description of this basic sociopolitical divergence, we begin in the nineteenth century (chapter 6). Then the historical evolution of these structures during previous centuries is examined, with examples taken from various parts of the mountain range, leading to a wide-ranging spatio-temporal discussion of the causes and consequences of different models of state formation (chapters 7 and 8). To assist the reader, most chapters in both parts of the book conclude with summaries of their main points, and chapter 9 synthesizes the key findings of the entire work.

The Alpine region with which we are concerned has an area of about 180,000 square kilometers. In a satellite photograph of Europe one can identify snow-covered surfaces and, in eastern areas, dense forests. Chiaroscuro shading indicates the beginning of the mountain arc next to the Ligurian Sea in the west, leading at first in a northerly direction, and then bending toward the east, stretching into central Europe. Its 1200-kilometer length separates Italy from northern parts of Europe (see illustration 1). Much of the Alpine region is over 2000 meters in altitude, and it reaches 4800 meters at the Mont Blanc, but it is also dissected by many low valleys—some of which are very low indeed. The Italian slope of the western Alps descends especially steeply onto the plain (see map 1.1). As is the case with many regions that are defined according to physical data, one can not identify boundaries for the Alps that are unequivocal. Any border between the Alps and the lower range of the Appennines in Italy is arbitrary, and the same is true for the delimitation between the Alps and the Dinarides in Slovenia (not indicated on the map).[6] Likewise, long stretches of the Alpine rim are difficult to differentiate from surrounding areas according to precisely consistent morphological characteristics. Still, when taken into account alongside the entire surface area of the Alps, these examples are relatively insignificant. For our purposes, defining precise boundaries is important only because statistical analysis requires a unified territorial basis. Therefore, for pragmatic reasons I will use two geographic

Chapter One

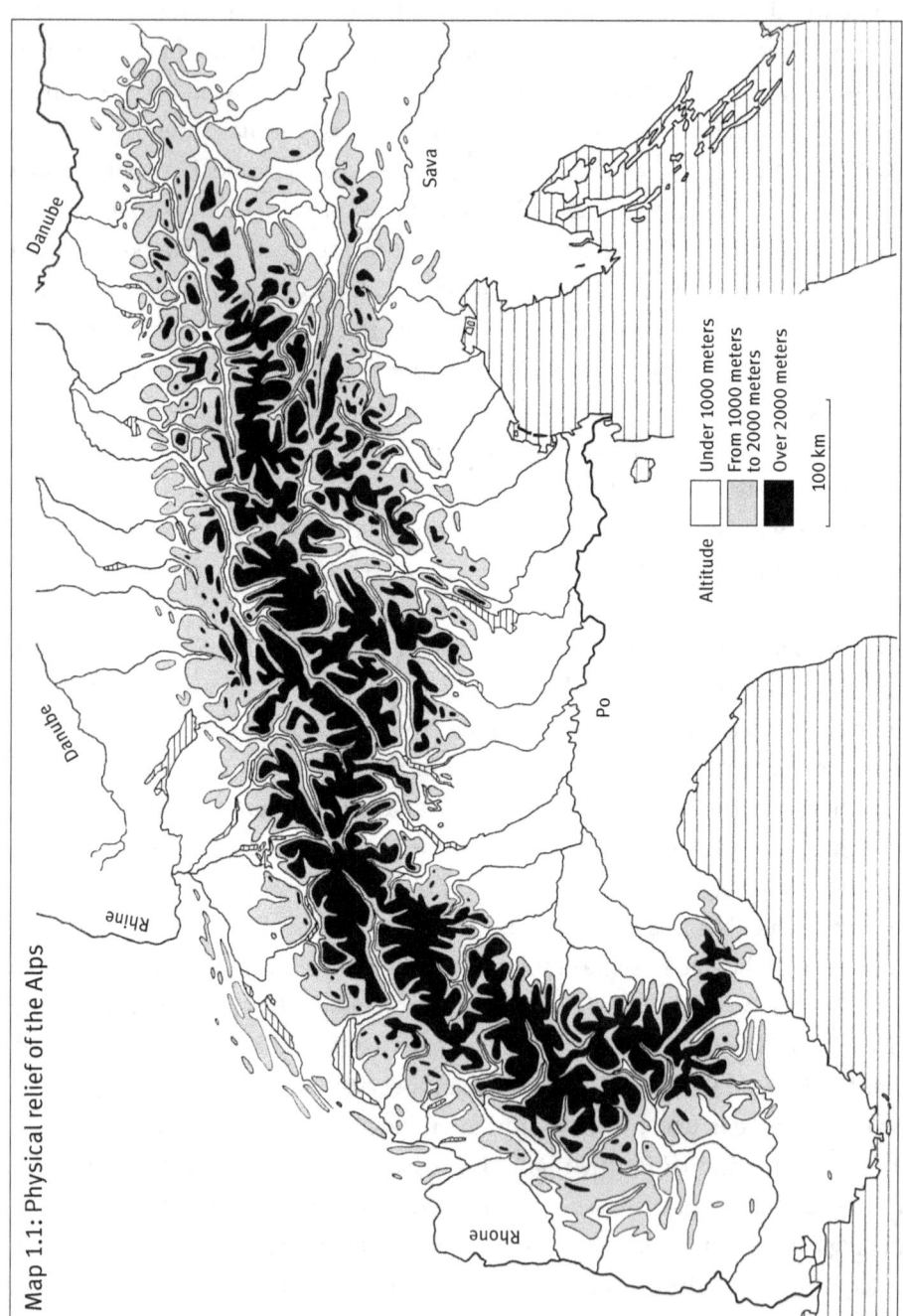

Map 1.1: Physical relief of the Alps

Source: Paul and Germaine Veyret, *Au cœur de l'Europe – les Alpes* (Paris, 1967), p. 28.

definitions—one more precise and one less so—since in historical research, it is often important *not* to identify precise boundaries.[7] This means deliberately including surrounding regions for comparative purposes, and systematically examining historical differences. This study will thus take into account conditions in surrounding areas as a matter of course.

One might ask why the four centuries between 1500 and 1900 have been singled out for study. Two interconnected phenomena explain the chronological point of departure. The beginning of the early modern period witnessed a burst of activity in the area of state formation and related changes. At the same time, for the sixteenth century onward, the availability of documentation, including quantitative data, is far greater than it had been for the late Middle Ages. The decision to continue the study up to 1900 was also made on the basis of source considerations: the era of statistics had its true beginning during the late nineteenth century. This makes it possible to obtain concrete economic and social indicators, and thereby to engage in international comparison—an especially valuable perspective when studying a politically heterogeneous space such as the Alps. New statistical possibilities also provide for the establishment of standards according to which individual historiographic traditions can be measured. Continuing from the early modern period through the end of the industrial nineteenth century also seems useful in that it avoids creating an artificial break between 'traditional' and 'modern' societies, underlining instead the processual character of historical developments.[8]

In the historiography that examines the Alps as a whole (to be discussed shortly), the centuries examined here are often represented as a period of decline. For some, the golden age of the Alpine world ended with the Middle Ages, for others with the eighteenth century. I am not primarily interested in distinguishing between positive and negative phases, but rather in sketching as broad a framework as possible, and one supported by statistics, in order to organize these arguments and assessments.

Quantitative history is sorely missing from Alpine research. The creation of quantitative data was linked to the birth of new processes of lordly and state control, which were limited in terms of territorial scope. Investigating a space such as the Alps thus means piecing together a multitude of evidentiary strands and making corresponding methodological judgments concerning the validity of the data.[9] With the exception of the quantitative analysis, which draws on both primary and secondary sources, this research is based mainly on the available published scholarly literature. The *corpus* of the national and regional historiography concerning the Alps (and surrounding areas)

Chapter One

is quite abundant for these four centuries. Given my own familiarity with the Swiss research, it has been an exciting experience for me to learn about the historiography from other, different parts of the Alpine arc that have little or no connection to each other. Although I have read widely in this literature, I remain quite aware of how far I am from a comprehensive overview, in the true sense of the phrase, and this suggests that there are serious obstacles to the possibility of writing an all-inclusive study.[10]

On the other hand, it is not too difficult to take stock of the scholarship dealing with the history of the entire Alpine realm. Apart from a few scattered precursors, this literature developed in tandem with the political and social initiatives described above, and is thus of relatively recent vintage. In my view, three works are especially significant and worth mentioning here.

Tout d'abord, Fernand Braudel. The first part of his famous study of *The Mediterranean and the Mediterranean World in the Age of Philip II* (published in 1949 and later revised and republished several times), focuses on the geographic environment, and in the first place, on mountain regions ("*Tout d'abord,* the mountains"). Braudel's opening salvo greatly stimulated later research on such areas. The mountain ranges around the Mediterranean are represented in a variety of perspectives in this first lively chapter, and the description is dappled with citations from archival sources and many details. A special role is attributed to the Alps, "exceptional mountains": unique because of their resources, their collective practices, the productive efficiency of their inhabitants, and their many roads. Braudel's panorama, presented in a literary style, both jump-started research in our field, and also provoked reactions. On the one hand, he brought mountain people into a historical narrative, and in so doing consolidated his concept of 'geographic time.' On the other hand, he also wrote them out of history: "The mountains are as a rule a world apart from civilizations, which are an urban and lowland achievement. Their history is to have none, to remain almost always on the fringe of the great waves of civilization, even the longest and most persistent."[11] These comments reflect Braudel the writer more so than Braudel the historian, and they have been often criticized of late for their unacceptably centralizing conception of history. Similarly, Braudel has also been criticized for having studied mountains from the outside, but not from the inside; and for having insisted on the fundamental importance of the impact of the environment on men, without also examining the opposite influence of men on the environment. In his preface to the third edition of this work (1976), Braudel distanced himself from his first version, drawing attention to his more recent, globally-oriented research. This new work alludes to new ways of approaching mountain regions, but it has not informed much other research in this area.[12]

Histoire et Civilisations des Alpes (1980) is a pioneering work edited by Paul Guichonnet that quickly became a classic. The very fact that the work's title explicitly links history and civilization and locates them in the Alps shows how different this work is from Braudel's. In Guichonnet one finds the questions "Où vont les Alpes?" and "Que faire des Alpes?" being asked of Europe itself, as the introduction makes clear.[13] Twelve contributors, almost all historians, take up these problems. In fact, this is the first work in which the history of the entire Alpine region, from prehistoric times to the present, is at the center of the discussion. With respect to the time period that concerns us here, the work's approach is complicated. First of all, the period is dealt with in several politically-focussed essays, broken down according to national areas. Other themes are discussed in a second series of pieces organized chronologically. Jean-François Bergier depicts the history of the period from the ninth through the sixteenth centuries as a cycle during the course of which the Alpine region evolves from a closed space (ninth through the eleventh centuries) into an open space (fourteenth and fifteenth centuries), and then back to a closed space with respect to relations with surrounding territories (fifteenth and sixteenth centuries).[14] Arnold Niederer provides an ethnological panorama of the "traditional" cultural and economic life of the Alps, that is, of the preindustrial structures that prevailed until the mid-nineteenth century, and often even later.[15] Finally, three authors describe more recent developments ending with the late twentieth century. The section dealing with the late nineteenth and early twentieth century discusses the development of the market, railway construction, tourism, and other issues, and is entitled "De l'autarcie à la dépendance."

During roughly the same moment at which geography was strengthening its own Alpine commitments and historians were on the verge of discovering this space themselves, anthropologists from around the world also became interested in the Alps. In particular, American anthropologists began their research on Alpine issues during the 1960s. Most of this research took the form of local studies situated in spatial and chronological contexts whose scale of analysis varied. Particular communities were analyzed in order to shed light on the salient theoretical debates of the day. Pier Paolo Viazzo's *Upland Communities: Environment, Population and Social Structure in the Alps Since the Sixteenth Century* (1989) originated as a local study too, but eventually shifted its attention to the entire Alpine space and its evolution over the long term. Of all of the authors mentioned here, Viazzo (as befits the specific tradition of his discipline) places the most emphasis on models that are explicitly formulated and discussed. His main interest is to study ecological anthropology by using the methodological and epistemological tools of historical demography. A central problem in this approach lies in determining the degree to which it is possible to analyze Alpine communities as closed systems that are completely reliant on their own local resources. Given the multiplicity of relations that one can observe, such as multiple examples of open systems in which migration and other external contacts played important roles (to cite one of the study's key findings), this perspective should neither be rejected out of hand, nor held to be unfailingly useful: "Much depended on the characteristics of local communal structures, and on their ability to resist economic and political pressure from outside."[16]

The basic queries of my book, and some of its arguments, are also drawn from other works, which are cited and discussed in following chapters. These Alpine-wide studies constitute an important foundation for the present work, setting the terms of the debate and providing a frame of reference that can not be ignored. Without a doubt, I have found myself most closely engaged by Viazzo's study. Precisely because of this closeness, clearly marked differences have also emerged. My work does not address the question of openness or closedness of village systems as a primary concern. Instead, it investigates the observable historical developments in both cases. In order to understand the economic dimensions of demographic processes, it is important also to examine them as factors driving change. Likewise, in order to grasp the sociopolitical dimensions of demographic change, we should examine the village context within the general process of state formation.

The political construction of territory

What is a historical space? A region with some sort of political cohesion, however it might have been created? Or a region whose inhabitants share certain experiences, without necessarily being aware of them? Or simply a region defined as such by historians who study it? Whether or not the Alps constitute a historical space depends on one's definition of the term. For the past few decades, the tendency has been to avoid taking for granted territorially-defined historical categories (such as 'the state'), emphasizing instead other kinds of territorial configurations. The dominant role that had obviously been played in historical narratives by the cohesion of the state has been lost. It still remains an important factor, though, since state formations certainly did have an impact on other sectors of society. Our analysis begins with a basic and brief

The Alps: A Historical Space?

overview of the various political formations in the Alpine territory, to set the stage for arguments developed later, coupled with a few preliminary observations.

According to an estimate made by Charles Tilly, Europe in the year 1500 or so comprised about 200 independent state formations. Four centuries later, just before the year 1900, there were 30. The expansion of the regions organized as territorial states or state-like formations is unmistakable, given this numerical decline. This quantitative change was accompanied by equally important qualitative changes. During the period studied, these societies were undergoing processes of territorialization and state centralization, causing one to wonder whether it really makes sense to try to count them. From the late Middle Ages onward, one incentive for state formation was the competition among different claimants to political power. Along with other phenomena contributing to the concentration of power, this rivalry increased over the course of the early modern period, exhibiting itself most clearly in the creation of standing armies. The need to pay for growing military expenses was another reason for the acquisition of new territories and the reinforcement of internal controls. In this fashion sovereigns augmented their authority, while at the same time the organization and coordination of groups and social interests increased. Courts, no longer peripatetic, were established in centers where administrative structures were concentrated, and whence they expanded. In federally-structured state formations, that is, in territories whose ties were not dynastic, the process of creating capital cities lasted until the nineteenth century. In the peripheries, vague and fragmented borders evolved progressively into linear boundaries, onto which the power of newly formed national states projected their might.[17]

Map 1.2 shows the approximate extension of the largest state formations (both sovereign and dependent ones), that existed in the Alpine and surrounding regions before 1790, together with the national boundaries that were delineated up to 1900. The abbreviations locate the residences or administrative centers of different power-holders. Even though smaller lordships, numerous during the eighteenth century, have been left out, the extraordinary diversity of the territories, both in terms of their internal structure and their external relations, is noticeable.[18]

- In the western Alps: *Provence* and *Dauphiné*, counties until the fourteenth and fifteenth centuries, when they were absorbed by the French crown. The high courts of justice, the fiscal courts, and the royal *intendants* (which appeared in the seventeenth century), were located at Aix and at Grenoble. As *pays d'états*, these two provinces, which stretched from the Rhône valley to the Alps, enjoyed greater autonomy than areas in central France, although their dependence on the center did in fact increase over time. With the French Revolution, in 1790, the provinces were divided into eight departments, which became the key administrative units.

- *Savoy-Piedmont,* to the north and east of the foregoing. The dignity of the house of Savoy's ancestral land was elevated from county to duchy during the fifteenth century, and in 1713 the dukes acquired a royal title. In the late Middle Ages the most important territories were located in the northern Alps, with their administrative center at Chambéry, indicated by SP on the map. After a period of foreign occupation, Emanuel Filibert transferred the ducal residence to Turin, shifting the nucleus of power to Piedmont. During the eighteenth century Savoy-Piedmont was reputed as a modern administrative state, and in the nineteenth century it became the leading edge of the movement to unify Italy. In 1860 the Savoyard territories were ceded to France.

- In contrast to the city-based republic of *Genoa,* the regions of *Lombardy* and the *Veneto* included extensive Alpine areas. Their territorialization originated during the late medieval period as the cities subdued their outlying lands, creating *contadi.* Later, smaller cities were forced to submit to the larger ones, and became integrated into their territorial states. After the Italian wars of the early sixteenth century, Milan and the rest of Lombardy fell under the domination of the Spanish branch of the Hapsburgs, and then in 1714 passed to the Austrian branch. The republic of Venice and its territory on the *Terraferma* remained independent until 1797. After the Napoleonic period, it also found itself under Austrian rule. Both regions were then unified with the rest of Italy as power passed from Vienna to Rome between 1859 and 1870.

- The *Swiss Confederation* was a matrix of alliances among thirteen rural and urban territories in the mountain zones and plateaus on the edges of the Holy Roman Empire. Subject territories held by the Confederation as a whole, to the north and south of the Alpine watershed, provided an element of cohesion. Until 1798 members of the Confederation met at diets held in various small locales (Schinznach, for instance, was not even a town, it was a small village with a hot spring – where the delegates could enjoy themselves in a variety of ways!) These meetings were at first presided by Zurich; later their presidency rotated. When the federal Swiss state was created in 1848, Berne was chosen as the seat of the new institutions.

- The *Valais* and the *Grisons* were two village-based republics that emerged from the episcopal lordships of Sion and Chur. They became part of the Swiss Confederation in the early nineteenth century, shortly after the Grisons's southern subject territory was joined to Lombardy as the province of Sondrio.
- The duchy and then kingdom of *Bavaria*, with Munich as its dynastic residence, held a narrow strip of mountainous territory; the two ecclesiastical states of the prince-bishops of *Trent* and *Salzburg*, on the other hand, were completely or almost entirely located within Alpine territory. Trent was partially a dependency of the Tyrol, to which it ceded parts of its lands on several occasions, eventually being incorporated in 1803. The territory of the archbishop of Salzburg was administered with greater independence and rigor. After its secularization, it shifted hands a number of times, and was then subsumed under the Austrian crown in 1849-50.
- Last but not least, the powerful house of Hapsburg. The European success of this dynasty is testified by its long possession of the imperial title and countless other signorial rights. At the beginning of the early modern period, Maximilian I unified the Hapsburg lands of the eastern Alps and surrounding lands, previously held by various branches of the dynasty. These duchies and counties became attached to a permanent center in Vienna over the course of the sixteenth and seventeenth centuries, with Ferdinand I and his successors. From 1564 until 1619–65 the house of Austria was newly subdivided, with cadet lines in Graz and Innsbruck. Provincial estates, especially the noble order, played a key role in the political development of the territory on the basis of the *Länder*. These assemblies met in the *Länder*'s administrative centers of Vienna and Linz (*Upper Austria*—above the Enns—and *Lower Austria*—under the Enns), Graz (*Styria*), Klagenfurt (*Carinthia*), Laibach/Ljubljana (*Carniola*), and Innsbruck (*Tyrol*).

Our compressed synthesis leaves out many finer details, necessarily overlooking many shifts in territorial and lordly configurations. Still, even a more detailed analysis would produce a similar result: since the beginning of the formation of modern states during the late Middle Ages, the power centers were mainly located at the margins or outside of the Alpine space. Among the eighteen different mountainous (or partially mountainous) state entities identified on map 1.2, two thirds did not have a dynastic residence or an assembly center located within the Alpine territory. Further, as previously mentioned, several of the six Alpine "centers" identified on the map

Chapter One

Map 1.2: Political divisions in the Alps, before 1790 and in 1900

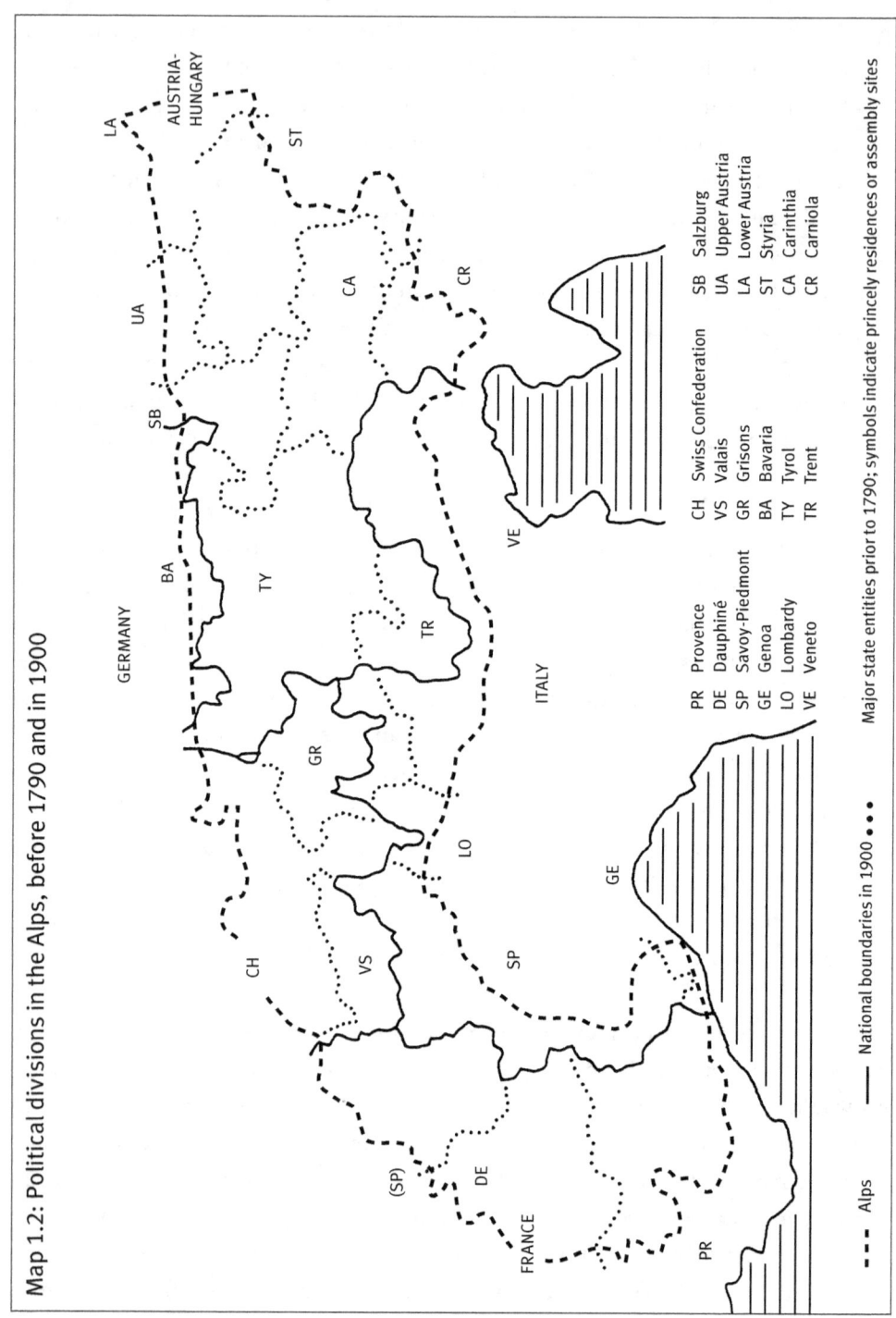

were more or less dependent on other places. This was the case for Innsbruck, which was temporarily a city of residence, and for Grenoble, the capital of Dauphiné. When considered apart from its relations with the French crown, Grenoble represents an interesting anomaly of political geography. It was the only clear example of a territory that included both Alpine and non-Alpine surrounding areas whose capital was located in the interior of the mountain range. The fact that Grenoble succeeded in its competition with the cities of the Rhône valley had an effect on urbanization, and will be discussed below.[19]

The distance of power centers from the Alps resulted in a significant level of regional and local autonomy. Examples can be found in various parts of the Alpine arc, from west to east, and on both the northern and southern slopes. The most obvious (and most often cited) formal example is provided by the mountainous regions of what would become Switzerland. But one should not forget that a large part of this area included subject territories that actually had a dependent status during the early modern period, when they were governed by lords appointed by urban or rural cantons. Still, the autonomy of these five small territories could really be far reaching: they were full members of the Swiss Confederation and enjoyed many sovereign rights. Their cohesion with the Confederation as a whole was only maintained by means of delegate assemblies. Localism in the Grisons, where about fifty jurisdictions and large communes were organized into three leagues, was at least as prominent. A clear political center was not established there until the nineteenth century; before then representative assemblies alternated between Chur and two other places.[20]

The tightening of cross-societal ties, more intensive administrative controls, and growing nationalism during the eighteenth and especially the nineteenth centuries had two consequences for the Alpine regions. They (1) brought power centers located in areas near the Alps closer to the mountains, and (2) increased the dependence of the Alpine regions on those centers. Examining the problem of borders provides a vivid example of this two-sided process. The nationalization of the Alps, on the one hand, created a sense of openness in which numerous local and regional divisions of every sort (cultural, economic, social, and political) were overcome. Small territories were opened up to much broader spaces in respective states. On the other hand, newly formed national borders became obstacles on a scale that was previously unknown, based on quasi-mystical state ideologies and on widespread growing militarization in the years before 1900.[21] The militarization of borders began early in the western Alps where there were relatively centralized states already in the late sixteenth century.

Chapter One

This process took a decisive turn in the eighteenth century, when France began the construction of a vast fortification system along its border with Savoy-Piedmont. During the period studied here, the western Alps also frequently witnessed important boundary changes. A valley like the Ubaye, today located in the department of the Alpes-de-Haute-Provence, was traded back and forth five times during the early modern period. The consequences of each new shift became heavier as the process of state building moved forward. Finally in the nineteenth century, all of Savoy was ceded to France (as mentioned above), and the "natural border" was established on the crest of the Alps.[22]

By 1900 the old, yet increasingly appealed-to theory of a natural border had turned geography into a political argument. Indeed, the theory was generated within politics, by the need to legitimate political objectives and outcomes. This had dangerous implications for those late nineteenth-century states whose lands straddled the Alps, such as the Swiss Confederation and the Tyrol (part of Austria). But soon there appeared a counter-theory—the idea of a "pass-state"—that turned to history in order to defend the notion that Alpine state formation had been focussed on the key transit routes, joining together regions on either side of inhospitable high mountain passes.[23] The sharpest battles within the Alpine zone took place in the Tyrol, where the claims of Italian irredentists clashed with those of Austrians and German nationalists. In 1919 the region was divided at the Brenner pass: the German-speaking *Südtirol*, now also called the Alto Adige, and the Italian-speaking Trentino, formerly referred to as the *Welschtirol* (Latin Tyrol), were acquired by Italy.[24] Fifty years later, the impulse to create ARGE ALP took off from Innsbruck, historic capital of northern Tyrol, uniting the Alto Adige/Südtirol and (soon thereafter) the Trentino together with other regions. ARGE ALP was for its part, as mentioned above, the first substantial institution of the regional Alpine movement.

However, this book investigates a different Alpine history, one that is less well-known and documented, because it reaches much farther back in time. The next chapter begins with the demographic evidence from the Alps, and then compares that evidence over the long term with data from surrounding areas.

Endnotes

1   NB: Complete references to selected studies relating to Alpine history and to the themes discussed here can be found in the bibliography; in the notes they are indicated by the name of the author(s) (or title) and publication date. "Convention sur la Protection des Alpes (Convention Alpine), Salzburg, 7 November 1991; Josette Barruet, Convention Alpine. Au-delà de l'effet catalyse," *Revue*

The Alps: A Historical Space?

de Géographie Alpine 83, 2 (1995): 113–121; unpublished constitution of the Working Community of the Alpine Regions; Bätzing 1991, 234–241; *CIPRA 1952–1992. Dokumente, Initiativen, Perspektiven. Für eine bessere Zukunft der Alpen*, ed. Internationale Alpenschutzkommission CIPRA (Vaduz, 1992), esp. 43–44.

2   Ibid.; *Die Zukunft der Alpen. Dokumentation ausgewählter Beiträge des Internationalen Symposiums "Die Zukunft der Alpen" vom 31. 8 bis 6. 9. 74 in Trento-Trient*, ed. Walter Danz, 2 vols. (Munich, 1975); *Le Alpi* 1974–75; Martinengo 1988, esp. 3–7, 597–609.

3   Dominik Siegrist et al., *Alpenglühn. Auf TransALPedes Spuren von Wien nach Nizza* (Zurich, 1993); from the regionalism literature, see Pierre Bourdieu, "L'identité et la représentation. Éléments pour une réflexion critique sur l'idée de région," *Actes de la recherche en sciences sociales* 35 (1980): 63–72; *Frontier Regions in Western Europe*, ed. Malcolm Anderson (London, 1983); Willy Erlwein, *Transnationale Kooperation im Alpenraum dargestellt am Beispiel der Arbeitsgruppe der Alpenländer (ARGEALP)* (Munich, 1981); for a case study of regionalism in northern Italy, which was particularly virulent beginning in around 1970, see *Lega e localismi in montagna. Il caso Belluno*, ed. Agostino Amantia and Ferruccio Vendramini (Belluno, 1994).

4   *Die Zukunft der Alpen*, 1: 149, and for the list of participants see 158–172.

5   See for example the conversation of Pierre Bourdieu with Lutz Raphael, "Über die Beziehungen zwischen Geschichte und Soziologie in Frankreich und Deutschland," *Geschichte und Gesellschaft* 22 (1996): 62–89; Reinhart Koselleck, "Erfahrungswandel und Methodenwechsel. Eine historisch-anthropologische Skizze," in *Alteuropa - Ancien Régime - Frühe Neuzeit. Probleme und Methoden der Forschung*, ed. Hans Erich Bödeker and Ernst Hinrichs (Stuttgart-Bad Cannstatt, 1991), 215–264.

6   See Bätzing 1991, 13; for the political-ideological elements of boundary-fixing, see Guglielmo Scaramellini, "Fra unità e varietà, continuità e fratture: percorsi di riflessione e ambiti di ricerca nello studio del popolamento alpino," in Coppola and Schiera 1991, 81–83.

7   See chapters 2 and 4 below.

8   On the periodization debate, see for example *Frühe Neuzeit - Frühe Moderne? Forschungen zur Vielschichtigkeit von Übergangsprozessen*, ed. Rudolf Vierhaus et al. (Göttingen, 1992).

9   International historical statistics have been based until now, for the most part, on collections of series of national data; in our case what is required is an international aggregation of regional data.

10 A general bibliography for the history of the entire Alpine arc does not exist, but the International Association for Alpine History, founded in 1995, has promoted initiatives of this sort: the first volumes of its annual journal *Histoire des Alpes* offers thematic bibliographies, and beginning with issue 10 (2005), one finds general advice for electronic bibliographic research in the Alpine space

11 Braudel 1972, 1: 25–53 ("Mountains Come First") and ibid., 1: 206–8 ("The Alps"); citation on 1: 34.

12 From among the rich literature on Braudel, one may consult Lutz Raphael, *Die Erben von Bloch und Febvre. "Annales"-Geschichtsschreibung und "nouvelle histoire" in Frankreich 1945–1980* (Stuttgart, 1994), esp. 109–137; Yves-Marie Bercé, "Préface," in Viallet 1993, 5–8; Jean-François Bergier, "Des Alpes traversées aux Alpes vécues," *Histoire des Alpes* 1 (1996): 12–13, 19; and Mathieu 1997, 124–125. For later reflections, see Fernand Braudel, *Civilisation matérielle, économie et capitalisme, XVe–XVIIIe siècles*, 3 vols. (Paris, 1979), for example 1: 46, 94, 127–29, and 3: 31.

Chapter One

13   The French phrase translates as "Where are the Alps going?" or "What to make of the Alps?"; Guichonnet 1980, 1: 10.

14   Jean-François Bergier, "Le cycle médiéval: des sociétés féodales aux Etats territoriaux," in Guichonnet 1980, 1: 163–264. Bergier began, at an early date, to examine issues dealing with the Alps as a whole; see the introductory essays in Körner and Walter 1996.

15   Arnold Niederer, "Economie et forme de vie traditionelles dans les Alpes," and "Mentalités et sensibilités," in Guichonnet 1980, 2: 5–136.

16   Viazzo 1989, 296; for the history of Alpine anthropology see esp. Albera 1995.

17   Tilly 1992, the figures are on 45–46; for an important assessment see Blockmans and Genet 1993; on the density of borders see for example *Grenzen und Raumvorstellungen (11.–20. Jahrhundert)*, ed. Guy P. Marchal (Zurich, 1996).

18   The sketch that follows is largely based on Imanuel Geiss, *Geschichte griffbereit*, vol. 4, *Schauplätze. Die geographische Dimension der Weltgeschichte* (Dortmund, 1993) (with bibliography).

19   Favier 1993; see chapter 4 below.

20   Hans Conrad Peyer, *Verfassungsgeschichte der alten Schweiz* (Zurich, 1978); Wolfgang-Amédée Liebeskind, "Altschweizerische Föderativsysteme," in id., *Institutions politiques et traditions nationales* (Geneva, 1973), 207–223; in general see also Pierangelo Schiera, "L'autonomia locale nell'area alpina. La prospettiva storica," in Schiera 1988, 3–50, 149–154.

21   Many Alpine studies of the 1970s and 1980s emphasize above all the new dependency and division of the Alps, a perspective that fails to reflect the ambivalence of the nationalization process.

22   Julien Coste, *Vallis montium. Histoire de la Vallée de Barcelonnette, Hautes terres de Provence, des origines à nos jours* (Gap, 1976), 41, 63–64, 99, 111–112; Guichonnet 1980, 1: 266–282, 296–302.

23   The detailed version, based on older works by Friedrich Ratzel, Aloys Schulte and other authors, can be found in Albrecht Haushofer, *Pass-Staaten in den Alpen* (Berlin-Grunewald, 1928); it is worth noting how little attention this discussion gave to actual historical traffic volumes.

24   Guichonnet 1980, 1: 399–412.

2 Population

As Massimo Livi Bacci has recently noted, modern population history is characterized by the discovery of "the demographic system." The upsurge, in around 1950, of research in historical demography led to subsequent efforts to isolate demographic variables from other factors and to insert them into systems models. During the same period other methodological approaches also emerged, including micro-demography and family reconstitution, the latter being a nominative method that proved very fruitful for fertility studies, but less useful for other kinds of demographic research. Since nominative methods are mainly based on series of restricted data sets, a related problem is how to generalize the results. For this reason, there have been, and continue to be, many calls for the integration of micro- and macro-demographic approaches. According to Livi Bacci there are other more important reasons for this integration: he argues that only through theoretically informed, exhaustive research of demographic systems is it possible to clarify their relationship to socioeconomic systems and to provide an adequate explanation of a population's behavior.[1]

However, if one's position is that demography is a subfield of history, then it is important to recognize that the opposite way of posing and explaining the question is equally significant. The demographic process can and should play its own role in helping to understand other kinds of phenomena. This is important particularly for the Alpine space, since the scholarly literature so often exaggerates environmental restrictions in

Chapter Two

a way that distorts genuine historical factors. The current state of the research in this field exhibits significant variation, on both national and regional levels. Of the studies that examine the entire Alpine arc, the most engaging is Pier Paolo Viazzo's *Upland Communities*. Viazzo's point of departure is his own historical demographic research in a village on the southern slope of Monte Rosa, and from there he shifts his anthropological attention to other Alpine territories, creating a compendium of results of similar research projects elsewhere. He also asks a series of new questions, obtaining a number of interesting results. Of particular importance is his conviction that the demographic regimes in the Alps were far more greatly differentiated than is usually assumed. But his research also reveals the kinds of problems entailed by the integration of micro- and macro-demographic approaches. This is immediately apparent in the data distribution, which is unavoidably non-systematic. For the sixteenth and seventeenth centuries, the absence of quantitative data is virtually complete; the micro-studies cited by Viazzo deal only with about twenty localities (out of the almost 6,000 communities in the Alpine space) for the period prior to 1900. The distribution of the macro-data is also random, and is not uniformly correlated to the mountains.[2]

The absence of spatio-temporal systematization, together with Viazzo's taste for experimentation, prompted me to look for other possibilities. This chapter limits itself to summarizing the empirical benchmark data produced by Alpine demographic history. It seemed important to me that these figures can claim to be rather representative. Since I also see the demographic process as an independent factor that helps us understand other historical phenomena, I will not attempt to explain it in detail. Rather, I will emphasize comparison with developments in the areas adjacent to the Alps. From a macro perspective, these comparisons can provide some starting points for the elaboration of a broader interpretation.

Data and collection methods

Most past endeavors to collect information about population levels stemmed from efforts to exert social control (and such is still the case). Regardless of how historical demographers situate themselves with respect to this fact, the statistics that they generate necessarily refer to politically-defined spaces. Yet the Alps are a geographically-defined space, which, in the course of history, have become a European border region, divided between various national states, each one with its own traditions of demographic statistics. Until quite recently, both current and historical data on the total population of the Alps have been very uneven. Paul Guichonnet, for example, numbered the inhabitants of the Alps in 1960 at 8.2 million, while another geographer calculated 10.9 million for 1950. The figures listed in table 2.1 do not even take into account the entire spectrum of the scholarly research on this topic,[3] and already represent a compromise: on the national levels, the numbers are even more scattered. Since the area at the foot of the Alps is densely inhabited in some places, the precise location of Alpine boundaries is crucial for this sort of data collection. This in turn has an impact on population estimates for pre- and proto-statistical time periods. Taking Guichonnet's figures for 1960 as a starting point, Jean-François Bergier first estimated the Alpine population in the year 1500 to be 1.5 million. Later, though, as a result of different assumptions about growth, he increased his estimate to 2 million. Naturally, had he used different figures for his point of departure, he would reached different estimates.

This chapter holds to Werner Bätzing's delimitation of Alpine space, as developed in a 1993 study whose consistent unit of analysis is the territory of individual communes. This permits one to define even large political regions in quantitative fashion, on the basis of a given percentage of Alpine area. As will be clear, this provides a methodological basis for historical statistical study. It is impossible to establish boundaries for the Alps without making subjective judgments. Bätzing's proposal functions within the usual parameters, but is less driven by political factors than is the definition of the Alpine Convention, which resulted from a clash of random interests whose effects led, to take one example, to a significant increase in the size of Germany's Alpine territory.[4]

Any attempt to examine the population of the entire Alpine arc prior to the beginning of modern census-taking retrospectively is doomed to face almost insuperable difficulties, especially if one works from the basis of communal data (Bätzing provides an estimate of this type for 1870). However, if one begins with data from politically unified regions, one can find relatively consistent data and estimates, or at least figures that can be interpolated, even for the early modern period. As has been shown, today's departments, provinces, cantons, and *Länder* are the best-suited units of analysis for a range of inquiries. It is important to remember, though, that these units were produced through different historical processes, and vary significantly in size. The differences in area between the small cantons of northern Alpine Switzerland and the sprawling Austrian *Länder*, for example, are immediately and strikingly apparent.[5]

For my demographic study, I have defined as Alpine regions those administrative units described above, as they existed in 1990, 75% of whose total area consists of Alpine land (according to Bätzing's definition). This is how the 26 territories designated on map 2.1 were identified.[6] As one can see from the map, this method effectively

Chapter Two

excludes many Alpine areas that are part of administrative regions whose territories also include large swaths of lowlands. Be that as it may, our selection still includes regions from every part of total Alpine space, 64% of which is accounted for in the selection (while 5% of the selection is Alpine-adjacent land). Below we will look at a few studies dealing with areas not included in our selection, in order to render our results more representative.

Table 2.1: Indications of total Alpine population levels, 1500–1900

Method / Author	Year	Population in millions
Retrospective estimate with 1960 as starting point, according to Guichonnet 1975 / Bergier 1980	1500	1.5
Retrospective estimate with 1960 as starting point, according to Guichonnet 1975 / Bergier 1988	1500	2.0
Presumed growth after 1500 / Bergier 1980	ca. 1750	1.5
Presumed growth after 1500 / Bergier 1988	1800–40	4.0
Territorial area according to Bätzing, communal data / Bätzing 1993	1870	7.0
Territorial area of the Alpine Convention, communal data / Bätzing 1993	1870	7.6
No area specifications, national data / Ruocco 1984	1871	7.5
No area specifications, national data / Ruocco 1984	1901	8.5
No area specifications, ethnic data / de Martonne 1926	ca. 1910	8.3
No area specifications, national data / Ruocco 1984	1951	10.9
No area specifications, national data / Guichonnet 1975	1960	8.2
Territorial area according to Glauert, national data / Glauert 1975	1970	12.3
No area specifications, national data / Ruocco 1984	1981	12.7
Territorial area according to Bätzing, communal data / Bätzing 1993	1990	11.0
Territorial area of the Alpine Convention, communal data / Bätzing 1993	1990	13.0

Territorial area: Alpine Convention—191,287 km²; Bätzing—181,489 km²; Glauert—180,000 km²

Sources, cited in abbreviated form: Bergier 1980 from *Histoire et Civilisations des Alpes*, ed. P. Guichonnet (Toulouse-Lausanne, 1980), 1: 175; Bergier 1988 from *Le Alpi per l'Europa. Una proposta politica* (Milan, 1988), 39; Bätzing 1993 from W. Bätzing, *Der sozio-ökonomische Strukturwandel des Alpenraumes im 20. Jahrhundert* (Berne, 1993), 47; Ruocco 1984 from *Les Alpes – ouvrage offert aux Membres du 25ᵉ Congrès International de Géographie* (Paris, 1984), 79; de Martonne 1926 from E. de Martonne, *Les Alpes. Géographie générale* (Paris, 1926), 124–25; Guichonnet 1975 from *Le Alpi e l'Europa. Atti del convegno di studi Milano 1973* (Bari, 1975), 2: 143; Glauert 1975 from G. Glauert, *Die Alpen, eine Einführung in die Landeskunde* (Kiel, 1975), 54.

Population

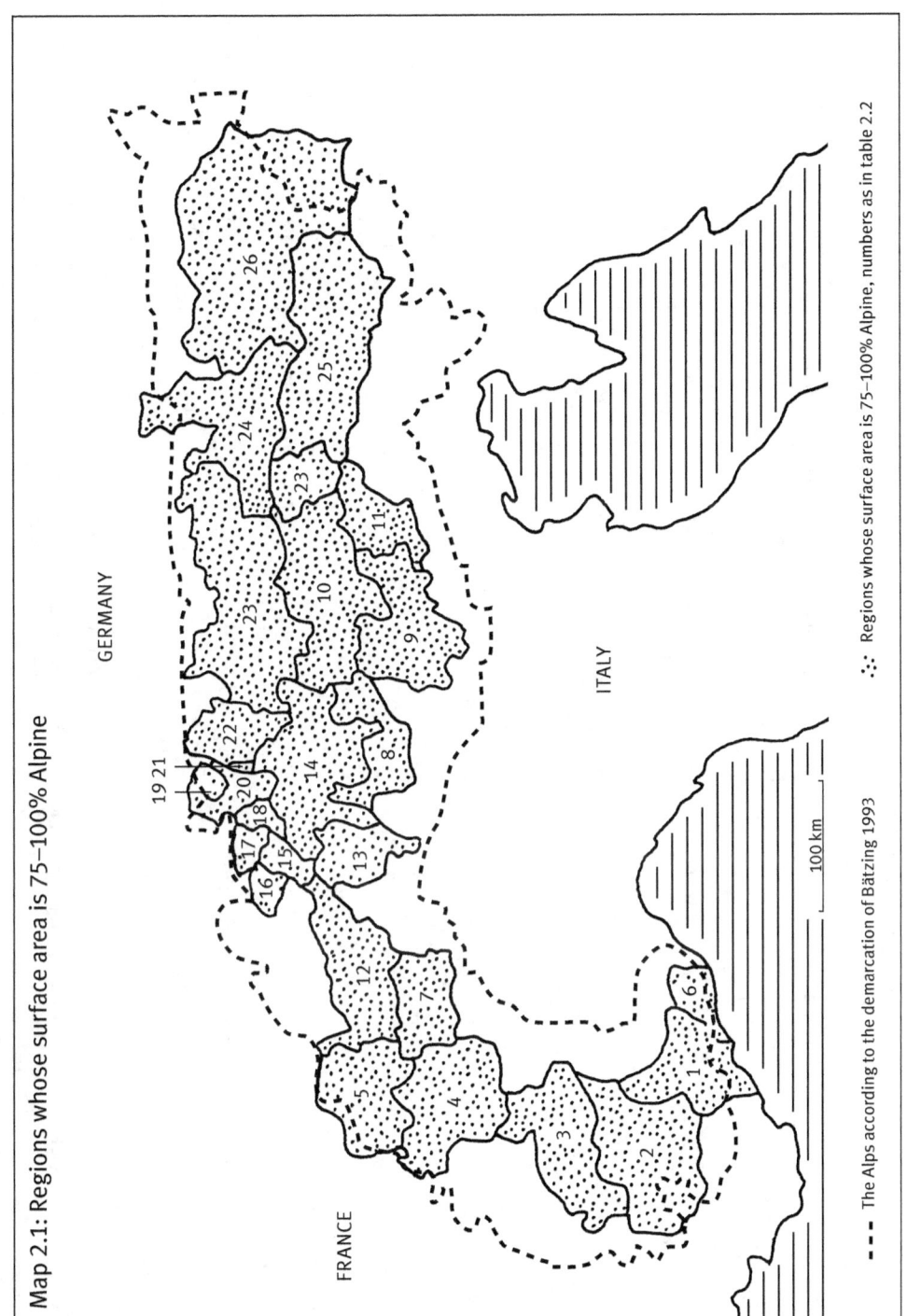

Map 2.1: Regions whose surface area is 75–100% Alpine

Chapter Two

Table 2.2: Population in Alpine regions, 1500–1900

Region		Population in thousands						Surface area (km²)	Alpine area as % of total
		1471–1543	1573–1615	1685–1723	1795–1810	1868–1872	1900–1901		
1	Alpes-Maritimes (F)	25.5	–	129.3	134.8	202.7	297.5	4299	87
2	Alpes-de-Haute-Provence (F)	35.4	–	130.5	134.0	139.3	115.0	6945	92
3	Hautes-Alpes (F)	44.1	–	94.8	112.5	118.9	109.5	5632	100
4	Savoie (F)	149.0	193.0	178.0	220.9	268.0	254.8	6230	100
5	Haute-Savoie (F)	137.0	179.0	152.0	184.1	273.0	263.8	4610	87
6	Imperia (I)	71.1	73.1	–	101.3	132.7	144.5	1155	75
7	Val d'Aosta (I)	–	100.0	61.9	64.9	84.1	84.2	3262	100
8	Sondrio (I)	68.0	98.0	82.4	86.0	117.4	131.0	3212	100
9	Trent (I)	–	167.0	171.8	227.0	335.0	355.0	6218	100
10	Bolzano/Bozen (I)	–	–	138.0	171.0	196.8	222.8	7400	100
11	Belluno (I)	48.2	69.3	92.9	108.4	196.0	220.7	3678	100
12	Valais (CH)	37.5	–	–	60.4	96.7	114.4	5226	100
13	Ticino (CH)	–	79.1	89.4	90.3	121.6	138.6	2811	100
14	Grisons (CH)	75.0	100.0	88.0	72.9	92.1	104.5	7106	100
15	Uri (CH)	–	8.8	9.6	11.8	16.1	19.7	1076	100
16	Unterwalden (CH)	–	–	13.7	19.1	26.1	28.3	766	100
17	Schwyz (CH)	9.6	14.3	21.0	34.2	47.7	55.4	908	100
18	Glarus (CH)	5.3	7.0	10.5	22.8	35.2	32.3	685	100
19	Appenzell (CH)	19.5	25.5	44.2	47.6	60.7	68.8	415	100
20	St. Gall (CH)	–	–	–	129.5	190.7	250.3	2014	79
21	Liechtenstein (FL)	–	2.7	–	5.0	7.5	7.5	160	100
22	Vorarlberg (A)	32.0	40.0	50.0	76.5	102.7	129.2	2601	100
23	Tyrol (A)	110.0	140.0	175.0	225.0	236.9	266.4	12,647	100
24	Salzburg (A)	75.0	100.0	125.0	141.0	153.2	192.8	7154	93
25	Carinthia (A)	135.0	175.0	215.0	267.6	315.6	343.5	9533	100
26	Styria (A)	279.0	320.0	380.0	512.3	721.0	889.0	16,387	79

Regions: Departments (F), Provinces (I), Cantons (CH), *Länder* (A), according to the administrative situation in 1990, listed from west to east and from south to north. Areas include unproductive land (in F without the subtractions made in the INSEE statistics). Percentages of Alpine surface areas are based on the parameters of Bätzing 1993, 39, calculated from his handwritten papers. The years and sources of the demographic data are listed in the appendix that follows.

Population

Appendix to table 2.2
This list indicates the years of the demographic data, the sources, and, for the numbers that were not directly transcribed, the estimation method employed. This information is listed in the format (region) years of demographic data / sources in the same order as the data, with complete citations for the first reference / estimate method and explanation. As the text indicates, one should not expect a high level of precision, especially for the data prior to 1800. For the following sources, which are used on several occasions, full citations are here:

INSEE Institut National de la Satistique et des Etudes Economiques, *Recensement général de la population 1990. Population légale. Fascicule départemental* (Paris, no year) (for regions 2–6: 1801, 1872, 1901).

ISTAT Istituto Nazionale di Statistica, *Popolazione residente e presente dei comuni. Censimenti dal 1861 al 1981* (Rome, 1985) (for regions 6–8, 11: 1871, 1901).

Bundesamt für Statistik, *Eidgenössische Volkszählung 1990. Bevölkerungsentwicklung 1850–1990. Die Bevölkerung der Gemeinden* (Berne, 1992) (for regions 12–20: 1798–1800, 1870, 1900).

(1) 1471, 1702–16, 1802, 1871, 1901 / Edouard Baratier, *La démographie provençale du XIIIe au XVIe siècle. Avec chiffres de comparaison pour le XVIIIe siècle* (Paris, 1961), 199–200; Gaston Imbert, *A la découverte d'une population, étude démographique des Alpes-Maritimes* (Aix-en-Provence, 1958), 17, 20; censuses of 1716 (Le Bret) and 1765 (Expilly), and communal data from Claude Motte, *Paroisses et communes de France. Dictionnaire d'histoire administrative et démographique* (Paris, forthcoming) Giuseppe Melano, *La popolazione di Torino e del Piemonte nel secolo XIX* (Turin, 1961), 4. / 1471: Comparison of hearths of 1702 with those of 1754–65 in three administrative districts; 1702–16 and 1871: integration of data missing in Imbert with the 1716 census and with Melano.

(2) 1471, 1716, 1801, 1872, 1901 / Baratier 1961, as in § (1), 164, 197–98; censuses of 1716 and 1765 as in § (1) / 1471: comparison of hearths with 1765 in eight administrative districts.

(3) 1472–76, 1706, 1801, 1872, 1901 / Joseph Roman, *Tableau Historique du Département des Hautes-Alpes. Etat Ecclésiastique, Administratif et Féodal antérieur à 1789* (Gap, 1993 [1887]), 169–71; Alfred Fierro, "La population du Dauphiné du XIVe au XXe siècle," *Annales de démographie historique* (1978): 360–401; Jean-Pierre

Brun, *Paroisses et communes de France. Dictionnaire d'histoire administrative et démographique. Hautes-Alpes* (Paris, 1995), 24–25, 32, 75–286. / 1472–76: comparison of hearths with 1699; data from Roman reduced by 8.5% based on Fierro. 1706: missing communal data taken from successive censuses; for Queyras data from 1698.

(4–5) 1500, 1600, 1700, 1801, 1872, 1901 / Roger Devos and Bernard Grosperrin, *La Savoie de la Réforme à la Révolution française* (Rennes, 1985), 34, 440; Raymond Rousseau, *La population de la Savoie jusqu'en 1861* (Paris, 1960), 124, 226–28; Dominique Barbero, *Paroisses et communes de France. Dictionnaire d'histoire administrative et démographique. Savoie, Haute-Savoie* (Paris, 1979–1980). / 1500 and 1600: according to the 1561 estimate by Devos (400,000) and the trend indications for the early and late sixteenth century indicated in Rousseau. Regression estimate 1801–1561 for 178 communes based on Barbero, used to estimate 1561 and for the distribution between the two departments.

(6) 1535, 1607, 1805, 1871, 1901 / Karl Julius Beloch, *Bevölkerungsgeschichte Italiens* (Berlin, 1961), 3: 303; Giuseppe Felloni, *Popolazione e sviluppo economico della Liguria nel secolo XIX* (Turin, 1961), 52. / 1607: based on the growth of 1535–1607 in the eastern part of the republic of Genoa.

(7) 1573, 1720, 1806, 1871, 1901 / Bernard Janin, *Une région alpine originale. Le Val d'Aoste. Tradition et renouveau* (Grenoble, 1968), 133; Beloch 1961, as in § (6), 270; Melano 1961 as in § (1), 34.

(8) 1500, 1596, 1697, 1797–98, 1871, 1901 / Guglielmo Scaramellini, "I rapporti fra le Tre Leghe, la Valtellina, Chiavenna e Bormio. Una difficile convivenza, una difficile valutazione," in *Storia dei Grigioni* (Bellinzona-Chur, 2000), 2: 154–57; Danilo Baratti, "La poplazione nella Svizzera italiana dell'antico regime," *Archivio Storico Ticinese* 111 (1992), 66; Guglielmo Scaramellini, *La Valtellina fra il XVIII e il XIX secolo. Ricerca di geografia storica* (Turin, 1978), 26. / 1596: the Grisons portion of the diocese of Como minus Poschiavo (estimated at 2000). 1697: Chiavenna estimated at 12,000. 1797–98: average of the two years.

(9) 1602, 1685, 1810, 1869, 1900 / Archivio Diocesano Tridentino, Visita ad Limina 1602, fol. 274r; ibid., 1685, fol. 110v; ibid., fol. 272r–274r (these data provided by Livio Sparapani); Hugo Penz, *Das Trentino. Entwicklung und räumliche Differenzierng der Bevölkerung und Wirtschaft Welschtirols* (Innsbruck, 1984), 83. / 1602 and 1685: regression estimate from provincial values for 1754, according to 1602–1685 growth, 1685–1760 in the diocese of Trent.

Population

(10) 1685, 1795, 1869, 1900 / Visita ad Limina 1685 and 1760 as in § (9); Bibliothek des Museums Ferdinandeum Innsbruck, Dipauliana 979/VIII (data provided by Kurt Klein); *Südtirol-Handbuch*, ed. Südtiroler Landesregierung (Bozen/Bolzano, 1991), 181. / 1685: regression estimate from 1754 according to 1685–1760 growth in the diocese of Trent.

(11) 1500, 1605, 1702, 1802, 1871, 1901 / Beloch 1961 as in § (6), 58–63, 163; *Relazioni dei rettori veneti in Terraferma*, vol. 2, *Podestaria e Capitanato di Belluno. Podestaria e Capitanato di Feltre* (Milan, 1974), xix; *Allgemeines Ortschaften-Verzeichniss der im Reichsrathe vertretenen Königreiche und Länder nach den Ergebnissen der Volkszählung vom 31. December 1900* (Vienna, 1902), 174. / 1500: regression estimate from 1548 based on rounded-off growth levels in three Terraferma districts. 1702: Belluno estimated at 36,250 on the basis of communicants in 1712; Cadore estimated at 20,600 according to growth rates in Feltre and Belluno from 1702 to 1802. All years: addition of 2.77% for Ampezzo-Buchenstein according to the censuses of 1900 and 1901.

(12) 1500, 1798–1800, 1870, 1900 / Pierre Dubuis, *Le jeu de la vie et de la mort. La population du Valais (XIVe–XVIe s.)* (Lausanne, 1994), 213–15; Leo Meyer, "Les recensements de la population du canton du Valais de 1798 à 1900," *Zeitschrift für schweizerische Statistik* 44 (1908): 292–307. / 1500: regression calculation based on 1798 with median growth rate of the existing data from seven communes for 1485–1524.

(13) 1597–1602, 1682–92, 1798–1800, 1870, 1900 / Baratti 1992, as in § (8), 61, 96. / 1682–92: parish of Brissago estimated at 2000.

(14) 1500, 1600, 1700, 1798–1800, 1870, 1900 / Jon Mathieu, La società rurale, in *Storia dei Grigioni* (Bellinzona-Chur, 2000), 2: 17.

(15) 1611–20, 1691–1700, 1798–1800, 1870, 1900 / Anselm Zurfluh, *Une population alpine dans la Confédération. Uri aux XVIIe–XVIIIe–XIXe siècles* (Paris, 1988), 528.

(16) 1700, 1798–1800, 1870, 1900 / Leonard Meister, *Kleine Reisen durch einige Schweizer-Cantone. Ein Auszug aus zerstreuten Briefen und Tagregistern* (Basel, 1782), 62–63; Urspeter Schelbert, *Bevölkerungsgeschichte der Schwyzer Pfarreien Freienbach und Wollerau im 18. Jahrhundert* (Zurich, 1989), 57. / 1700: regression calculation from 1743 based on growth rates in Schwyz 1700–1743.

(17) 1500, 1600, 1700, 1798–1800, 1870, 1900 / Schelbert 1989, as in § (16), 51.

Chapter Two

(18) 1543, 1576, 1700, 1798–1800, 1870, 1900 / Hans Rudolf Stauffacher, *Herrschaft und Landsgemeinde. Die Machtelite in Evangelisch-Glarus vor und nach der Helvetischen Revolution* (Glarus, 1989), 271.

(19) 1535, 1597, 1713–34, 1798–1800, 1870, 1900 / Hanspeter Ruesch, *Lebensverhältnisse in einem frühen schweizerischen Industriegebiet. Sozialgeschichtliche Studie über die Gemeinden Trogen, Rehetobel, Wald, Gais, Speicher und Wolfhalden des Kantons Appenzell Ausserrhoden im 18. und frühen 19. Jahrhundert* (Basel-Stuttgart, 1979), I: 207–9; Walter Schläpfer, *Wirtschaftsgeschichte des Kantons Appenzell Ausserrhoden bis 1939* (Gais, 1984), 136–39; Markus Schürmann, *Bevölkerung, Wirtschaft und Gesellschaft in Appenzell Innerrhoden im 18. und frühen 19. Jahrhundert* (Appenzell, 1974), 34, 55. / 1535: average value of divergent data in Ruesch. 1597 and 1713–34: Ausserrhoden according to Schläpfer; *Inneres Land* of Innerrhoden according to Schürmann; *Äusseres Land* of Innerrhoden estimated at 20% of Innerrhoden.

(20) 1798–1800, 1870, 1900

(21) 1584, 1806, 1868, 1901 / Alois Ospelt, "Wirtschaftsgeschichte des Fürstentums Liechtenstein im 19. Jahrhundert. Von den napoleonischen Kriegen bis zum Ausbruch des Ersten Weltkrieges," *Jahrbuch des Historischen Vereins für das Fürstentum Liechtenstein* 72 (1972), appendix 25–27; *Statistisches Jahrbuch 1991 Fürstentum Liechtenstein*, ed. Amt für Volkswirtschaft (Vaduz, 1991), 20. / 1584: low figure based on the number of houses and on a comparison with the Vorarlberg.

(22–26) 1527, 1600, 1700, 1800, 1869, 1900 / Kurt Klein, "Die Bevölkerung Österreichs vom Beginn des 16. bis zur Mitte des 18. Jahrhunderts," in *Beiträge zur Bevölkerungs- und Sozialgeschichte Österreichs*, ed. Heimold Helczmanovszki (Vienna, 1973), 105. / The chronologically standardized data were used.

Table 2.2 lists demographic data from the 26 Alpine regions for each century between 1500 and 1900, with additional information for the years around 1870, in order to link up with Bätzing's communal data. The sources and methodological observations are located in the appendix to the table. Generally, the procedure and the data quality can be characterized as follows.

As far as possible, population numbers have been taken from existing studies. When confronted with divergent figures, preference has been given to those taken from official statistics (in particular for the years around 1800), or else to those obtained from studies that appear to be more trustworthy—otherwise, the table reports the average of the divergent values.[7] As indicated earlier, the numbers provided by available publications vary widely from region to region and country to county. The simplest situation is presented by Austria, for which Kurt Klein's reliable 1973 study calculated the population levels of all of the Austrian *Länder*, according to consistent criteria, beginning in the early sixteenth century. For the period prior to the first systematic census of 1754, estimates are given whose margin of error is 5–10%, according to Klein. Klein did not revise his data in later studies, and they have been accepted by most scholars.[8] For Italy and Switzerland, unlike Austria, the state of the research varies from region to region. It is extremely difficult to obtain early figures for the French departments, not so much because of the state of the records, which is often favorable, as because scholars have tended to focus on kinds of demographic problems other than overall population levels. Additionally, the new administrative subdivisions introduced during the French Revolution created a deep divide between the *ancien régime* and the nineteenth century. In order to evaluate long-term developments, regions have to be standardized according to today's territorial divisions. This requires the redistribution of the pre-Revolutionary population into the departments created as a result of the Revolution. Although here this seems particularly anachronistic, the method seeks only to create a spatially coordinated system, "un découpage commode de l'espace" (Bernard Lepetit).[9]

The primary sources for these figures are just as diversified as the secondary ones. In some regions, census-taking began early: in the province of Belluno, a Venetian territory, the earliest records of this type date from 1548; in Savoy they appear in 1561. During the eighteenth century, when genuine statistical interests begin to assume real weight alongside state interests in census-taking, systematic data collection was carried out in many regions. Still, the Alpine area discussed here was not completely covered until about 1800. Even though some of these collection projects were remarkable for their time, they cannot be compared to the national censuses carried out around 1870, and still less with those of ca. 1900.[10] Nonetheless, this category of registration generally provides more reliable results than the numerous documents that refer to population levels only indirectly and for a variety of reasons (by counting hearths, houses, numbers of inhabitants eligible to bear arms, communicants, etc.), and thus require problematic mathematical calculations to obtain total population levels.[11]

The calculations in which I engaged to produce my figures were mainly related to territorial standardization. After the First World War, for example, the province of

Chapter Two

Belluno grew through the addition of the Tyrolese districts of Ampezzo-Livinallongo/Buchenstein, which in 1900–1901 accounted for scarcely 3% of the provincial population. This percentage was used to adjust pre-1900 figures for this region. In most of my estimates, though, the synchronic dimension is less important than the diachronic one. I use regression methods if there is evidence of growth during certain periods and if earlier data is fairly incomplete, and I occasionally base growth indicators on data from only portions of the territories considered, or even from surrounding regions (though in such cases I proceed even more cautiously). A basic requirement, though one that can only be partially satisfied, is that the units of analysis represented by these data (houses, communicants, etc.) are the same.[12]

As the regression estimates indicate, the unreliability of the data, as a rule, increases with chronological distance. Many of the regional figures for the sixteenth and seventeenth centuries should be taken as hypotheses. But we can also be sure that at the very least we are dealing with hypotheses that were formulated by a series of different authors. One of the main tasks of research is to re-verify such numbers continuously. Historical criticism is especially useful when it can offer well-founded and reliable ways of doing this. Finally, with respect to temporal standardization: given multiple factors of uncertainty, I have hazarded few estimates, accepting instead substantial chronological gaps. For the data around the year 1900, collection times differ in terms of months, and these gaps grow gradually to 72 years for the fifteenth and sixteenth centuries. Much of the data are dispersed in smaller chronological periods, though, as indicated in the table's appendix. It is risky to try to establish long-term tendencies on the basis of limited data during moments of rapid demographic change. This is the case here especially for the growth period during the late fifteenth and early sixteenth centuries. However, attention has been given to making sure that the sample analysis contains regional data before and after 1500.

Comparing long-term trends

The population of the Alps appears to have almost tripled between 1500 and 1900, as indicated by a comparison of the sums of the first (1471–1543) and sixth (1900–1901) columns of table 2.2. To produce a total estimate in absolute terms, we use Bätzing's figure for 1870 as a baseline for comparison and calculate the numbers for each region according to the minimum loss method. Approximate results indicate that the population of the Alps according to the definition described above was 2.9 million around 1500, 4.0 million around 1600, 4.4 million around 1700, 5.3 million around 1800, and 7.9 million around 1900. The rates of annual increase for each of the four centuries would be 3.2, 1.0, 1.9, and 4.0 ‰.[13]

The variation in these Alpine trends run parallel to those established for the population history of Europe as a whole. Still, it makes sense to identify a more limited sub-European sector for purposes of comparison. We have chosen countries that today contain some portion of Alpine land: France, Italy, Switzerland, Austria, Germany, and Slovenia. The annual growth rates in this group of countries (within their current borders with roughly equal margins of error) for the four centuries in question are approximately 3.9, 0.5, 3.2, and 5.9 ‰. Differences with the rates in the Alpine space thus appear to have grown significantly over the long term. During a first phase, up to 1700, differences were minimal, but in the following two centuries, the difference in rates grew to 1.3 ‰ and then as high as 1.9 ‰. Still, since the population density in this group of countries was substantially greater than in the Alpine space, even at the beginning of the period in question (about 25 people/km² as opposed to 16 people/km²), the absolute difference of density increased noticeably even when differences in growth rates were small.[14]

Historical demography cannot ignore the relationship between population and space,[15] a relationship which—in the Alps—is tied to a number of imponderable factors. As is obvious, the mountains include vast areas of so-called unproductive land. But neither the dimensions of these areas, nor their level of productivity is precisely known. Figures provided by national statistics vary significantly, both within a single country over time, and from country to country at the same moment. The ways in which national statistics represent productivity (or lack thereof) is closely tied to the attitude of the day held by national administrations. What is less well-known, or at least often ignored by scholars, is the fact that real surface area is not measured, but only projected surface area. This means that the area-size of given sector of steep terrain is artificially shrunk. It has been estimated that if the real surface area of the Alpine region were evenly flattened out, it would measure 240,000 km² rather than 180,000 km². This difference would be greater than the total area of unproductive terrain (however this might be defined), but it also contains a fictional element, in that, generally speaking, the steeper a given terrain, the less productive it is. The mutual counteraction between the factors of steep terrain and unproductivity reinforces our preference for using the simple measure of projected surface areas.[16] Village altitude is a more useful indicator of agrarian potential and productivity. Altitude is an easier measure to use (one can identify the village center as a reference point), and is closely related to the potential for agricultural use.[17] Before returning to the comparative discussion of Alpine and surrounding regions, we will now consider the various developmental tendencies within the Alpine space.

Table 2.3 presents data relating to population growth and density in the 26 Alpine territories. On a macro-level, this information shows that demographic developments in the mountains varied significantly. Given the fragility of our database, it makes sense to use caution in interpreting the data for the early phase, though for the seventeenth century one may reasonably be more confident in the accuracy of the growth numbers. Five of the regions, grouped in two areas, display indications of noteworthy demographic decline during this crisis-stricken period.

Various scholars have attributed this decline to the epidemics that took innumerable lives during the 1630s. In the Val d'Aosta, the plague of 1629–31 was "incontestablement un immense désastre." According to an official report of 1646, two thirds of the population disappeared, in the town of Aosta not even a dozen homes were spared, and large portions of the valley were completely depopulated. Comparative evidence is available for nearby Savoy, one of the Francophone territories in which the series of epidemics from 1630 to 1637 was especially devastating. The military operations of the Thirty Years' War aggravated the crisis both here and in the second of the areas marked by demographic decline. In the province of Sondrio, records from before and after the plague outbreak show population losses far superior to those of the Ticino (which at the time belonged to the same diocese as the Valtellina). Losses were also great in the Grisons just to the north, though in this region the greatest decline occurred during the eighteenth century, in southern districts, due to a massive emigration of artisans and merchants. This is one piece of evidence pointing to an above-average impact of migratory movements on demographic phenomena. Prior to 1900, economically-active migrants from the Grisons could be documented in almost 600 European towns. During the nineteenth century, similar migration was one of the main reasons for the negative population trends in the southern French Alps (departments of Alpes-de-Haute-Provence and Hautes-Alpes).[18]

Population

Table 2.3: Population growth and density in Alpine regions, 1500–1900

Region	Annual growth rate in ‰				Persons per square kilometer					Area (km²)	Avg. altitude (m)
	16th c.	17th c.	18th c.	19th c.	~1500	~1600	~1700	~1800	~1900		
1 Alpes-Maritimes (F)	6.8	6.8	0.4	8.0	6	—	30	31	69	4299	566
2 Alpes-de-Haute-Provence (F)	5.3	5.3	0.3	−1.5	5	—	19	19	17	6945	773
3 Hautes-Alpes (F)	3.3	3.3	1.8	−0.3	8	—	17	20	19	5632	999
4 Savoie (F)	2.6	−0.8	2.1	1.4	24	31	29	35	41	6230	642
5 Haute-Savoie (F)	2.7	−1.6	1.9	3.6	30	39	33	40	57	4610	632
6 Imperia (I)	0.4	1.6	1.6	3.7	62	63	—	88	125	1155	272
7 Val d'Aosta (I)	—	−3.3	0.6	2.7	—	31	19	20	26	3262	951
8 Sondrio (I)	3.8	−1.7	0.4	4.1	21	31	26	27	41	3212	596
9 Trent (I)	—	0.3	2.2	5.0	—	27	28	37	57	6218	698
10 Bolzano/Bozen (I)	—	—	2.0	2.5	—	—	19	23	30	7400	850
11 Belluno (I)	3.5	3.0	1.5	7.2	13	19	25	29	60	3678	751
12 Valais (CH)	1.6	1.6	1.6	6.3	7	—	—	12	22	5226	1004
13 Ticino (CH)	—	1.4	0.1	4.3	—	28	32	32	49	2811	550
14 Grisons (CH)	2.9	−1.3	−1.9	3.6	11	14	12	10	15	7106	1129
15 Uri (CH)	—	1.1	2.0	5.1	—	8	9	11	18	1076	784
16 Unterwalden (CH)	—	—	3.4	3.9	—	—	18	25	37	766	539
17 Schwyz (CH)	4.0	3.9	4.9	4.8	11	16	23	38	61	908	628
18 Glarus (CH)	8.5	3.3	7.9	3.5	8	10	15	33	47	685	604
19 Appenzell (CH)	4.3	4.4	1.0	3.7	47	61	107	115	166	415	825
20 St. Gall (CH)	—	—	—	6.5	—	—	—	64	124	2014	573
21 Liechtenstein (FL)	—	2.8	2.8	4.3	—	17	—	31	47	160	547
22 Vorarlberg (A)	3.1	2.2	4.3	5.3	12	15	19	29	50	2601	696
23 Tyrol (A)	3.3	2.2	2.5	1.7	9	11	14	18	21	12647	880
24 Salzburg (A)	3.9	2.2	1.2	3.1	10	14	17	20	27	7154	736
25 Carinthia (A)	3.6	2.1	2.2	2.5	14	18	23	28	36	9533	648
26 Styria (A)	1.9	1.7	3.0	5.5	17	20	23	31	54	16387	542

Population growth and density according to table 2.2 and its appendix. The figures refer to the data contained therein, which for the most part do not correspond exactly to the precise century-year breaks. The growth rates *in italics* refer to intervals greater than a century without intermediate data. Altitudes are averages of communal centers according to national statistics (F 1988, I 1988, CH 1952, FL 1991, A 1981).

Table 2.4: Population growth and density in the Alps and in surrounding areas, 1500–1900

Country Region	Annual growth rate in ‰				Persons per square kilometer					Area (km²)	Alpine area as % of total
	16th c.	17th c.	18th c.	19th c.	~1500	~1600	~1700	~1800	~1900		
Countries:											
1 France (F)	3.7	1.6	2.6	3.3	24	35	41	53	74	551208	7
2 Italy (I)	2.9	0.1	3.0	6.3	33	44	44	60	112	301277	17
3 Switzerland (CH)	4.8	2.8	3.4	6.8	14	22	29	40	80	41293	65
4 Austria (A)	2.5	1.5	3.8	6.7	18	21	25	37	72	83856	65
5 Germany (D)	5.9	–1.4	4.5	8.7	21	37	32	50	121	356804	2
6 Slovenia (SLO)	—	—	2.0	4.9	—	—	33	41	65	20251	34
Regions:											
7 (F) three Alpine departments	*5.1*	*5.1*	0.8	3.1	6	—	21	23	31	16876	93
8 (F) three departments in surrounding areas	*6.9*	*6.9*	1.8	6.3	7	—	40	47	87	14645	14
9 (I) three Alpine provinces	3.6	0.4	1.8	5.2	18	26	26	32	54	13108	100
10 (I) two regions in surrounding areas	4.4	–0.1	3.5	6.8	43	66	66	93	185	35306	27
11 (CH) six Alpine cantons	3.3	1.1	0.6	4.0	13	18	20	22	32	13001	100
12 (CH) *Kornland* in surrounding areas	6.4	4.3	4.2	7.6	19	36	55	84	179	14318	0
13 (A) three Alpine *Länder*	3.6	2.2	2.1	2.4	11	14	18	22	27	29334	98
14 (A) three *Länder* in surrounding areas	2.1	1.3	4.5	8.2	25	29	33	52	119	35117	28

"Alpine" = Alpine regions with a maximum of 15% of non-Alpine surface area. "Surrounding areas" = contiguous regions with a maximum of 30% of Alpine surface area.

The regions are:
(7) Departments of the Alpes-Maritimes, Alpes-de-Haute-Provence, Hautes-Alpes. (8) Departments of the Bouches-du-Rhône, Var, Vaucluse. (9) Provinces of Belluno, Sondrio, Trent. (10) Regions of Lombardy and Veneto, without the Alpine provinces of Sondrio and Belluno. (11) Cantons of Appenzell, Glarus, Grisons, Schwyz, Ticino, Uri. (12) Non-Alpine Switzerland, surface area according to Bätzing 1993, 39. (13) Federal *Länder* of Carinthia, Styria, Tyrol. (14) *Länder* of Burgenland, Lower Austria, Upper Austria. Administrative arrangements of 1990.

Surface areas of countries according to the encyclopedia Brockhaus, 1986–94 edition (F without the subtractions made in the INSEE statistics). Percentages of Alpine area taken from table 2.2. The demographic data do not always correspond exactly to the precise century-year breaks. The growth rates *in italics* refer to intervals greater than a century without intermediate data. The years and the sources for the population growth and density are indicated in n. 24.

Table 2.3 also shows that there were already notable gaps, at the start of the early modern period, between the population densities of various Alpine regions. The differences at the extreme western end of the Alpine arc are partly a function of the year from which the records date. The very low figures for the French regions come from the 1470s, while the unusually high ones from the nearby Italian province of Imperia date from 1535. During the intervening years, a number of localities indicate a phase of prominent growth. The small province of Imperia, located at a low altitude, 25% of whose territory is non-Alpine (as defined by this study), was undoubtedly densely populated. The Ligurian coast was among the most densely populated regions of Italy. Appenzell was another densely populated Alpine region located on the edge of the mountains. It is almost the smallest region in our sample group, only 400 km² in size, rendering it somewhat problematic as a point of comparison. Still, in the seventeenth century its population increased very quickly, and in the following century the canton was held to be one of the most densely population rural regions in Europe. "If I had not witnessed these things myself," remarked one traveller, "I would never have believed that a district of this size could contain so many inhabitants." The high population density was both a cause and a consequence of the precocious concentration of artisanal activity and industrialization that revolutionized the foundation of the region's economy between the sixteenth and nineteenth centuries, as several studies have shown.[19] In 1900 only about 20% of the total active population was involved in agriculture.

The only other places with such low percentages of agricultural workers were the nearby cantons of St. Gall and Glarus: in contrast, half of the regions studied here had percentages above 60%. But the percentage of agricultural workers is a complicated and imprecise quantity. It refers not only to the amount of agricultural activity as a portion of all economic activity, but also to the level of agricultural specialization (since the production of material goods within farming complexes should count as agricultural activity but is unrecorded). The imprecision derived mainly from the failure to taken into account women's activities. Since statistical practices varied from country to country, and sometimes from region to region, detailed comparisons become problematic, even if one considers other indicators.[20]

In general terms, though, one can not discount the existence of a relationship between population density and percentages of agricultural workers. At the turn of the twentieth century, densely-inhabited regions like Appenzell and St. Gall often counted a low percentage of farm workers, who were more numerous in thinly-

Chapter Two

inhabited regions such as the Valais, the Alpes-de-Haute-Provence, and the Hautes-Alpes. But there were also contrary cases: the provinces of Imperia and Belluno, the Trentino, and the Haute-Savoie were all characterized both by medium to high population density and by high percentages of farm workers. The province of Sondrio should also be mentioned: it was slightly less densely inhabited, but counted 82% of its population as active in agriculture—the highest percentage of our sample group. This is an important warning against underestimating the possibilities of agricultural development in the Alpine space. When examined from a different perspective, the data illustrate the overall importance and dynamics of Alpine agriculture during our period. In 1870, the percentages of workers involved in agriculture had been considerably higher than in 1900; the differential for the Alpine regions was 10% on average. It is plausible that pre-1870 numbers would have been even higher. According to the data presented, though, the population had grown significantly since the beginning of the early modern period. Together, these two observations lead to the conclusion that over the course of the period the expansion of agricultural production had been noteworthy.[21]

How did altitudinal location affect population growth? The average regional altitude figures listed in table 2.3 are the result of a massive simplification that is only justified by our method of examining entire regions. Let us first limit ourselves to the consistent data set for the Austrian regions. For the sixteenth and seventeenth centuries, population growth was not limited by altitude—the data almost illustrate the opposite—whereas this relationship does emerge, more clearly with time, during the eighteenth and nineteenth centuries. This is especially obvious for Styria, a region of low altitude that is partly outside of the Alpine area. During the first two centuries of our period, this region experienced the slowest growth, but in the nineteenth century it surpassed every other region. At the same time the Tyrol, the highest region in altitude and one that hitherto experienced medium growth levels, fell to the bottom.

Table 2.3 reports the growth rates for 22 of the 26 Alpine regions between 1700 and 1900. If we divide the data according to average altitude, creating two groups of regions under or over 750 meters, we see clear differences during the eighteenth and nineteenth centuries. The average growth rates of low-altitude regions increased from 2.2 to 4.1 ‰ during those centuries, while those of higher regions only amounted to 1.5 and 2.7 ‰. The average population densities of these two groups already differed by 1700 (23 versus 19 persons/km^2). But the most significant differences came later, with 31 vs. 20 persons/km^2 in 1800 and 49 vs. 21 persons/km^2 in 1900.[22] During the study period, altitude, as a factor of regional population growth, seems to have become increasingly important over time. Still, this does not mean that other factors lost their importance. If one takes altitude into account, the Austrian regions, such as Styria, were among the least populated in around 1800, and the same is true for the Alpes-de-Haute-Provence in around 1900.[23]

Such regional differences are normal for European demographic history. They existed in lowland areas as well, and on a large scale, as one can see from the upper section of table 2.4, which reports the data available for present-day countries with Alpine territories. To facilitate a comparison, in terms of averages, between the Alps and surrounding land, the lower section of table 2.4 contains four groups of regions. These groupings were organized according to various factors: availability of data, ratio of Alpine area to total size of territory, and location within the Alpine space. Although it was not possible to satisfy completely any single criterium (for example, population estimates for the sixteenth century had to be synthesized), on the whole this grouping should provide a useful comparative frame.[24]

The situation may be summarized as follows. With respect to the entire period between 1500 and 1900, the population increased in all of the Alpine regions more slowly than in the surrounding regions. During the sixteenth and seventeenth centuries, though, there were territories (in the east-central section, and especially in the northeastern—Italian and Austrian—portion of the Alpine arc) whose mountainous areas were characterized by growth rates superior to those in the lowlands. Since the population densities in the Alps and in surrounding areas already differed at the beginning of the study period, not even more rapid population growth could narrow the absolute difference in density between the two. In the eighteenth and nineteenth centuries, the general divide in growth patterns widened, becoming the rule for all of the groups, further increasing the gap in density levels as a consequence.

By way of conclusion, we will look briefly at those Alpine regions not included, for methodological reasons, in our sample regions. In the Dauphiné there is a 3304 km²-district within the present-day French department of Isère, whose population increased between 1476 and 1800 somewhat more rapidly than in a smaller, less mountainous territory elsewhere in the region for which comparable data exists. After 1800 though, the trend reversed itself. René Favier has studied the Dauphiné as a whole, and found that, as early as the eighteenth century, growth rates varied between *bas* and *haut Dauphiné*. The Piedmontese side of the western Alps was more densely populated than the French side; Raoul Blanchard has analyzed this from the perspective of altitudinal differences. In the *hautes vallées* the population grew between 1734 and 1838 by an annual rate of 3.5 ‰. Soon thereafter, until 1901, the population declined, the rate dropping to -2.6 ‰. In the *basses vallées* the rate was higher especially during the second period, such that the entire population of the Piedmontese mountains increased slightly over the course of our study period. Markus Mattmüller has found that in the canton of Berne, which stretches deep into the Alpine space (with an area of 3876 km^2 according to the definition used here), the population increased more rapidly in the Alpine *Oberland* than in the lower-lying *Mittelland* between 1499 and 1764 (a 6.6 ‰ increase in the former zone as opposed to a 5.2 ‰ increase in the latter). In the nineteenth century Alpine growth remained high, but was now exceeded by growth in the lowlands. The provinces of Udine and Pordenone (with 4137 km^2 of mountainous area) experienced, between 1548 and 1802, levels of population growth that were strikingly superior to averages in the lowlands (2.8 vs. 0.6 ‰); the nineteenth century brought with it both accelerated growth and a clear reversal of the trends in each area.[25]

Despite the necessary caveats that must be made here, including with respect to these data, the examples reinforce the impression that during the earlier period, the impact of a mountainous location and high altitude on demographic development was quite limited, but that the influence of these factors became much more important during the later portion of the study period. In other words, one can consider this impact to be the result of an historical process.

To summarize: a study of Alpine population history on a vast scale first requires a method that will allow us to depict a geographically-defined space with the help of politically-defined units of analysis that lend themselves to demographic evaluation. Taking a cue from Werner Bätzing's 1993 outline of Alpine space, which used communes as constituent units, we have defined a representative study area of 26 Alpine regions. These regions are administrative districts of national states. Their size and traditions of political stability enable a reconstruction of their demographics, reaching back to the start of the early modern period. The regions included in the study area are those whose surface area is 75–100 % Alpine. Data collected from published scholarship or estimated by us for each century between 1500 and 1900 indicate that over the course of the period the Alpine population almost tripled. Within the area defined according to the parameters outlined above, the population in 1500 was about

2.9 million, and in 1900 it was around 7.9 million.

The multiformity of this demographic process was apparent in the first instance within the Alpine space itself, on the regional level. Analysis according to altitudinal differences shows that factors linked to altitude became increasingly important over the course of the study period. Similarly, when one compares Alpine regions with surrounding regions, and the Alpine space as a whole to the total areas of the present-day states containing Alpine regions, one finds a parallel trend. In both cases the differences in demographic growth rates were initially small, even sometimes favoring the mountains, and then these differences grew rapidly in the eighteenth and nineteenth centuries. Still, since the areas surrounding the Alps had higher population densities, the difference in densities between Alpine and surrounding areas rarely narrowed in absolute terms, even when growth rates were higher in the Alps.

Most of the documents that can be used to reconstruct demographic processes were generated in the interest of social control; for this reason they constitute an element in the field of social forces. In the Alpine space in particular, political dynamics quickly transformed observations about demographic levels into polemics about over- or under-population—polemics in which assertions could count for more than reasoned findings.[26] The long term has seen an enormous increase in both the production of statistics and public access to them, but their use has always remained controversial. Time periods marked by general fears of overpopulation, for instance, were not exempt from competing arguments. By the late nineteenth and early twentieth centuries, the upper hand in the Alps appears to have been won by those who complained about depopulation, lamenting for example the "flight from the highlands." It is true that some localities and regions experienced both increasing population density and population loss. In general, the difference between Alpine population levels and those in surrounding areas was increasing. But these complaints about depopulation were also being made at a moment when the Alpine lands had just completed a century of more rapid growth than during any of the earlier periods studied here.

Endnotes

1 Livi Bacci 1997.
2 Viazzo 1989 (with an extensive bibliography).
3 A *Habilitationsschrift* completed in 1987 estimated the then-current population of the Alps at 7.5 million, and that of forty years earlier at "about 3 million"; see Hans Gebhardt, *Industrie im Alpenraum. Alpine Wirtschaftsentwicklung zwischen Aussenorientierung und endogenem Potential* (Stuttgart, 1990), 33. It is a bit ironic that, further below, we will suggest a similar figure—but for more than

Chapter Two

four hundred instead of forty years ago.

4   Bätzing 1993, 24–45.

5   In the emerging statistical system of the European Union, the Austrian *Länder* are, significantly, situated at a higher classificatory level than the other regions under consideration here; see Martin Schuler, "Le système statistique européen," *La Vie économique* 67, 10 (1994): 30–38.

6   In Germany (with 3% of the Alpine area) and Slovenia (with 4%) there are no territorial units of a comparable area that justify their inclusion; the principality of Liechtenstein is here treated as a region.

7   Sharp differences are found in Savoy (according to a controversial 1960 study by Rousseau; see appendix to table 2.2) and in the canton of Appenzell (according to an only partly plausible 1979 revision by Ruesch; see ibid.)

8   See for example Kurt Klein, "Bevölkerung und Siedlung," in *Geschichte Salzburgs* 1983–1991, 2: 1289–1360.

9   A fundamental and content-rich work for France provides few indications of population levels for the earlier periods, and organizes the eighteenth-century population into regions based on the *Généralités*; see Dupâquier 1988 (as in n. 24), 2: 75–78.

10   Between about 1890 and 1910, the quality of the censuses seems to have been above average, even by today's standards; see for example CH-Statistics 1990, xviii.

11   Detailed descriptions of the sources and the problems posed by them may be found in the appendix to table 2.2.

12   A 1961 study by Baratier (see § 1 of appendix to table 2.2), compares the *larem foventes*—hearths—of late fifteenth-century Provence with the inhabited houses there during the eighteenth century. On a definitional level these units appear to be almost the same, but their size was variable. For this reason, Baratier does not quantify the inhabitants. Here a more explicit form of the same comparison will be made, resulting in an estimate of the population based on the eighteenth-century situation.

13   Our point of departure value for 1870 is 6,963,226 (Bätzing 1993, 47). The numbers cited are derived by comparing the regional figures that are occasionally available with the 1870 number, and the year of the estimate is calculated as a median. If one follows the regional values back in time, step by step, one has to accept larger losses of data, but the oscillations should become clearer. In this way, the numbers of inhabitants in millions would be 3.0, 4.1, 4.4, 5.3, and 7.9, and the annual growth rates would be 3.1, 0.7, 1.9, and 4.0 per thousand.

14   According to the data listed in table 2.4 and referenced in n. 24 below; Slovenia was estimated for 1500 and 1600 according to the increase in Styria and Carinthia. In around 1700 the population density was about 38 vs. 24 persons/km^2, and in 1900 it was 95 vs. 44 persons/km^2.

15   Livi Bacci and other researchers have argued against research tendencies that limit themselves to a purely demographic interpretive approach; see Livi Bacci 1997, 150–151.

16   Sandgruber 1978, 35–36; Günter Glauert, *Die Alpen, eine Einführung in die Landeskunde* (Kiel, 1975), 13; Bätzing 1993, 40.

17   There are many indications that recent changes in communal structures, given the large number of communes, have little bearing on the results of this research. In Switzerland, the 1952 altitudinal data have been used, since these figures do not appear in later territorial statistics.

18   Janin 1968, 133; Giuseppe Prato, "Censimenti e popolazione in Piemonte nei secoli XVI, XVII

e XVIII," *Rivista Italiana di Sociologia* 10 (1906): 337; Dupâquier 1988 (as in n. 24), 2: 197–206; Scaramellini 1999 (see § 8 of appendix to table 2.2); Baratti 1992, 63–68 (see ibid.); Mathieu 1992, 90–107; id., "Migrationen im mittleren Alpenraum, 15.–19. Jahrhundert. Ein Literaturbericht," *Bündner Monatsblatt* (1994): 347–362; Anne-Marie Granet-Abisset, *La route réinventée. Les migrations des Queyrassins aux XIXe et XXe siècles* (Grenoble, 1994), esp. 93–100.

19   Ruesch 1979, 78, citation from W. Coxe (see § 19 of appendix to table 2.2); Albert Tanner, *Spulen - Weben - Sticken. Die Industrialisierung in Appenzell Ausserrhoden* (Zurich, 1982).

20   The percentages of agricultural workers from 1870 to 1900 were collected by means of reconciling different national categories and according to two criteria (with and without women); an important source of regional variation with respect to the numbers of women counted is gender-specific mobility. The relationship to density is calculated here from the median between the two criteria; the data are in the appendix (table A.1).

21   Depending on the region and the period, other activities, especially mining, played in important role; see Klein 1973, 75, 85–86, 95.

22   All of data refer to the median regions considered for each time period. The average gives a lot of weight to the large Austrian regions and shows greater disparity in growth rates between the two categories (2.2/1.3 and 4.3/2.6‰), and a smaller increase in population density disparities (25/17, 31/20, 48/25 persons/km^2).

23   Measured according to the classification of average altitude and population density; the small Swiss regions display a high level of dispersion.

24   For each region of table 2.4 the following list includes: the years of the demographic data / the sources, with a full reference when first cited / the method followed for the estimates themselves. In order to minimize data loss, for three Alpine regions I made estimates that pre-date the time frames indicated in table 2.2. In order to calculate growth rates in regions with chronologically dispersed data, the median was set as the normative year.

(1)   1500, 1600, 1700, 1801, 1901 / Jacques Dupâquier, *Histoire de la population française* (Paris, 1988), 1: 382, 2: 68, 3: 123 / 1500: regressive calculation from 1560 according to the lowest growth rate for the period 1450–1560.

(2)   1500, 1600, 1700, 1800, 1900 / Athos Bellettini, *La popolazione italiana. Un profilo storico* (Turin, 1987), 14, 40.

(3)   1500, 1600, 1700, 1798, 1900 / Markus Mattmüller et al., *Bevölkerungsgeschichte der Schweiz*, part I (Basel and Frankfurt a.M., 1987), 1: 4, 365.

(4)   1527, 1600, 1700, 1800, 1900 / Kurt Klein, "Die Bevölkerung Österreichs vom Beginn des 16. bis zur Mitte des 18. Jahrhunderts," in *Beiträge zur Bevölkerungs- und Sozialgeschichte Österreichs*, ed. Heimold Helczmanovszki (Vienna, 1973), 105.

(5)   1500, 1600, 1700, 1800, 1900 / Christian Pfister, *Bevölkerungsgeschichte und historische Demographie 1500–1800*, vol. 28 of *Enzyklopädie deutscher Geschichte* (Munich, 1994), 10, 19–22; Colin McEvedy and Richard Jones, *Atlas of World Population History* (Harmondsworth, 1978), 69 / 1500–1700: calculation based on the surface area of 1990 according to the demographic situation of 1800.

(6)   1700, 1818, 1910 / *Enciklopedija Slovenije*, vol. 9 (Ljubljana, 1995), 250–251 / 1700: regressive calculation from 1754 according to the growth in Carinthia and Styria.

(7)   1471–74, 1706–16, 1801–02, 1901 / see table 2.2, appendix (1–3).

Chapter Two

(8) 1471, 1709–16, 1801, 1901 / Édouard Baratier, *La démographie provençale du XIIIe au XVIe siècle. Avec chiffres de comparaison pour le XVIIIe siècle* (Paris, 1961), 195–196; censuses of 1716 (Le Bret) and 1765 (Expilly), communal data provided by Claude Motte, *Paroisses et communes de France. Dictionnaire d'histoire administrative et démographique* (Paris, forthcoming); Roland Sicard, *Paroisses et communes de France. Dictionnaire d'histoire administrative et démographique. Vaucluse* (Paris, 1987); F-Statistics 1990 / see table 2.2, appendix (1–2).

(9) 1500, 1596–1605, 1685–1702, 1797–1810, 1900–1 / see table 2.2, appendix (8, 9, 11) / Trent 1500: regressive calculation from 1602 according to the median growth of Sondrio and Belluno.

(10) 1500, 1600, 1700, 1800, 1900 / Karl Julius Beloch, *Bevölkerungsgeschichte Italiens*, vol. 3 (Berlin, 1961), 163–164, 242, 353; I-Statistics 1871, vol. 1, xvi; I-Statistics 1985, 2–6 / Regressive estimate from 1871 onward according to the growth in the various regional former states.

(11) 1500–1543, 1576–1615, 1687–1723, 1798–1800, 1900 / see table 2.2, appendix (13–15, 17–19). / Ticino and Uri 1500: regressive calculation from 1600 and 1615 according to the growth in Sondrio and Schwyz, respectively.

(12) 1500, 1600, 1700, 1800, 1900 / Mattmüller 1987, as in (3) above, vol. 1, 356–364; id., "Agricoltura e popolazione nelle Alpi centrali, 1500–1800," in *Le Alpi per l'Europa. Una proposta politica*, ed. Edoardo Martinengo (Milano, 1988), 65.

(13–14) 1527, 1600, 1700, 1800, 1900 / Klein 1973, see (4) above, 105.

25 Fierro 1978, see table 2.2, appendix (3), 359 (Grésivaudan with 3304 km^2 and Viennois-la-Tour, left bank of Rhône, with 1170 km^2); Favier 1993, 42–44, 465; Blanchard 1938–1956, 6: 321, 326, 334–37; a comparative presentation for the western Alps, esp. for the nineteenth and early twentieth centuries, can be found in ibid., 7: 519–588; Mattmüller 1987 (as in n. 24), 131; CH-Statistics 1990, 4 (the figures for the nineteenth century are estimated from 7 districts in the *Oberland* and 10 in the *Mittelland*)—this may be compared to Pfister 1995; Paolo Fortunati, "La Popolazione Friulana dal secolo XVI ai giorni nostri," in *Atti del Congresso Internazionale per gli Studi sulla Popolazione*, ed. Corrado Gini, vol. 1 (Rome, 1933), 118–119; see also Daniele Beltrami, *Forze di lavoro e proprietà fondiaria nelle campagne venete dei secoli XVII e XVIII* (Venice, 1961).

26 Already in the late sixteenth century, Venetian governors in the mountainous region of Friuli reported devastating population losses, though this phenomenon is not reflected in the demographic data that they submitted; Beloch 1961, see table 2.2, appendix (6), 45.

3 Agriculture and Alpiculture

"There are no grounds less susceptible of improvement than mountainous pastures," wrote Thomas R. Malthus in 1803. "They must necessarily be left chiefly to nature; and when they have been adequately stocked with cattle, little more can be done." This assessment of Swiss mountain areas seemed of utmost importance to him, because he believed Alpine pastures to be the centerpoint of Alpine agriculture. He also saw in their limited environmental potential an illustration of his general theory of population. From different perspectives and with different intentions, before and after his "Essay on the Principle of Population," many voices have pointed out the limited resources of the Alpine space. At least since the beginning of the early modern period, mountain inhabitants have often drawn attention to the infertility of their lands when required to justify themselves publicly (in fiscal matters, for example); and from the nineteenth century onward, when modern agrarian science began to estimate the development prospects of large areas, Alpine regions did not fare favorably.[1] Nonetheless, between 1500 and 1900 the total population of the Alpine space probably almost tripled, and there is no doubt that this growth was linked to a notable increase in agrarian production. How was this kind of growth possible?

Chapter Three

This chapter takes up this question by examining different important aspects of Alpine agriculture. Our point of departure is not a happy one: although the past decades have seen a tendency to study the history of the Alpine arc as a whole, there are few overviews of the agrarian history of this European frontier area. A large portion of the specialized literature is either nationally or regionally oriented, or it is primarily anthropological or geographic in nature. Agrarian histories of single countries analyze Alpine regions therein in quite different ways, partly because of the differing percentages of a state's total area represented by those regions. For example, while in Austria the Alpine regions play an important role and in Italy they are often taken into consideration, in the French syntheses they are referred to only sporadically.[2] The necessity for Alpine historiography to avail itself of discussions with contiguous disciplines is demonstrated by the fact that the most detailed, complete exposition of agriculture in the past comes from human geography. In 1940 and 1941, John Frödin published a two-volume work on *Zentraleuropas Alpwirtschaft* (*The Alpiculture of Central Europe*), in which he established the results of many years of research, partly from a historical perspective. We will return to this work.[3]

The relationship between Alpine environment, population, and agriculture can be examined from a variety of perspectives. In what follows, I begin with the assumptions that a growing population required new and more frugal ways of exploiting the environment, and that these innovations were influenced not only by new consumption needs, but also by labor inputs. In a pre-industrial context, space-saving methods of land exploitation usually required intensive labor inputs, and thus depended on certain population levels. Often, increased output did not match the amount of extra labor required; older, extensive methods of farming thus enjoyed higher labor productivity and remained in place as long as a territory's resources permitted.[4] This development model considers the environmental potential within a total historical context, unlike the perspective of many geographic studies. The latter, because of their disciplinary affiliation, tend to examine nature's impact on Alpine agrarian structures rather than looking at other correlations. This privileges, *a priori*, environmental conditions as explanatory factors.[5]

The notion that agrarian labor could be especially burdensome in mountain regions is often reaffirmed, though usually in general terms. Our investigation would require more precise information, in quantitative form if possible, concerning various forms of agrarian activity in different circumstances. The state of the documentation and of the research leaves much to be desired in this regard. Still, there are both different methods for estimating labor inputs and occasional numerical data. For example, for technological contexts such as those that predominated prior to 1900, it has been estimated that in the French Alps a field of one hectare required 230 man-hours of labor per year to produce hay; for other crops the numbers were as follows: cereals 315, potatoes 2500, and wine 2750. From this one can deduce that, given the same amount of territory, forage production required three-quarters as much labor as cereals, and that potatoes (which had to be hoed) required a huge effort (8 kg of cereals were produced per hour of labor, compared to 5 kg of potatoes). We will see below that economic parameters varied greatly in time and space, limiting the claims that single examples permit us to make.[6]

Let us begin by looking at the importance of alpiculture. Then we will examine the forms of land use, the intensification of livestock-raising and plant cultivation. Finally the technical component of agricultural evolution will be taken into consideration.

The intensity differential in the Alps

Among the different branches of Alpine agriculture, alpiculture—the economy of the Alpine pasture—is prominent in the scholarly literature. It is sometimes identified in historical overviews as the primary source of agricultural income, it is often highlighted in local and regional monographs, and it is the subject of a body of specialized literature. An easy definition of alpiculture as an economic form is hard to come by, since the examples are so varied. Among its most important and diffuse characteristics are summertime pasturing, the elevation of the pasture areas, and the mode of exploitation—carried out at a distance from settlements, but juridically and economically dependent on them. The preeminent position of alpiculture within the scholarly tradition is less obvious than one might assume. In fact, this position resulted not only from concern for the real conditions in Alpine space, but also from comparison to areas surrounding the Alps. To place this in the correct perspective, one must take into account the fact that the face of agriculture in the surrounding areas had changed in many ways between the sixteenth and the nineteenth centuries.

Changes in northern Italy, often referred to by travellers as the "garden of Europe," were particularly rapid. Around 1500 the area south of the Alps was densely populated for the time, and marked by intensive agriculture that proceeded in some areas without letting the land lie fallow. Until around 1800, land was farmed almost everywhere with annual or sub-annual rotations. Pastures and wooded areas shrank in size during the agrarian intensification of the early modern period. During the eighteenth century, the

Chapter Three

size of the permanently cultivated area in Italy is estimated to have increased by 10%, and probably even more in northern Italy. At the same time, the number of irrigation and land improvement projects multiplied, and new farming and livestock-raising practices were introduced. These included corn-growing, which spread widely after its appearance in the sixteenth century; extending rice cultivation across the region; increasing forage production by introducing and rotating new plants; and more frequent stabling of animals.[7]

These processes became even more dynamic in the nineteenth century, consolidating the long-standing differences with the nearby Alpine space. Territorial statistical data for extensively-exploited lands remained rather imprecise through the end of the period under study, but regional differences still emerge clearly from figures such as those recorded in lands subject to Hapsburg domination. In 1861, in the kingdom of Lombardy-Veneto (which included some mountainous regions), 33% of the productive territory was registered as wooded or pasture land. In the Tyrolean lands (including the Vorarlberg, the northern Tyrol, the southern Tyrol/Alto Adige, and the Trentino), the Salzburger lands, and Carinthia, the same proportion was 72–75%. In the northern Hapsburg lands of Upper and Lower Austria, most of which were outside the Alpine space, the figure was 41%. Cadastral records also permit a view of change over time. For example, the percentage of pasture land in Lower Austria diminished between 1789 and 1897 from 9.4% to 3.7%. Over the same period, in the Alpine territories cited above, pasture land was estimated at a constant 25–40% of the total.[8]

The first reports on alpiculture should be read against this background of change. The decline over the long term in the amount of resources used extensively had several consequences. On one level, it generated new interest in remote mountain pastures. This interest was in turn refracted by related developments such as the commercial success of certain Alpine products and the Rousseauian return to 'nature.' Ludwig Wallrath Medicus, a German writer, published a description of Swiss alpiculture in 1795 that took as its starting point the idea, then widespread, that any increase in crop production in the plains presupposed an increase in livestock-raising. More careful management of the livestock sector was thus necessary, and according to Medicus, alpiculture (ignored in agricultural handbooks) could be a source of valuable knowledge. He imagined alpiculture first of all as economically intensive. "My reaction to the fertility of the Alpine pastures resembled that of other travellers: I had heard so much talk about the fruitfulness and productivity of these places that I pictured cows grazing in grass as high as their bellies, because I mistakenly imagined these mountain pastures to be cultivated meadows." Halfway through his journey, Medicus was greatly struck by the real Alpine pasture lands and by their "very short grass," which nonetheless made possible an economy based on livestock and cheese-making, activities that promised a "secure subsistence, even sometimes significant wealth."[9]

As time passed, scholarly resources and perspectives broadened. The Swedish scholar of human geography John Frödin began to dedicate himself to the study of alpiculture during the 1920s, when he could rely on a large scholarly literature, some of which was statistically informed, on the subject. Field research was equally important for his work, and he carried this out across the entire Alpine arc, from Carinthia to the southern French Alps. Unlike Medicus, he looked at alpiculture from the point of view of its extensivity, because he was interested in comparing it to other kinds of pastoral economies. By differentiating alpiculture from nomadism and transhumance, he also included in it the fields and meadows that tied the population to a fixed residence and provided livestock with the forage required for winter stabling. In this wide sense of cultural comparison, "alpiculture" meant virtually the same thing as "agrarian mountain economy." Frödin stressed the importance of Alpine land use by referring to its geographic diffusion and its deep history, though he was unable to provide much historical material for earlier periods. Most important, within his broad concept of alpiculture, he took the significance of the Alpine pasture lands for granted, without engaging in serious study.[10]

As these classic works show, the basic assumptions adopted in the course of scholarly analysis are extremely important and should be clarified at the beginning. Alpiculture can be considered from three perspectives: as a particular branch of mountain agriculture in relation to the plains (e.g. Medicus); as an extensive form of exploitation in relation to other forms of the pastoral economy (e.g. Frödin); or, as a component of a regional context of mountain economies. This third perspective is crucial for our problematic, and we intend to discuss it briefly, on the basis of an example from a distinctly Alpine region.

If one examines the first set of reasonably reliable statistics from the Grisons, collected around 1900, one finds that the 800+ Alpine pastures registered were occupied by livestock for an average of 87 days per year. In some cases the average was significantly lower, and in other cases it was higher. The sources for the early modern period, scarce for the earlier centuries but more frequent after 1750, indicate that this datum can be taken as an approximation for our entire study period. We can thus affirm that for three quarters of the year, livestock was not to be found in Alpine pastures, but was fed with the resources of lands at lower altitudes, either in pastures during the periods of transition, or with forage during the long winter. These lower slopes produced not only hay but also cereals, and, depending on the valley, also chestnuts, fruit, and wine. The number of people and workers who could be supported on the Alpine pastures during the summer months varied according to the structure in place. In the Grisons, it appears that the prevalent model was that of a cheese-making consortium in which salaried workers took care of the animals and made cheese. Local documentation indicates that these workers were only a very small percentage of the community's total population. And what can be said of dairy production? From the late eighteenth century onward, economic studies are in agreement that the total alpicultural production of butter, cheese, and ricotta was markedly inferior to that of the valleys. The summertime output of a single cow in many places during that time amounted to about a third of its annual output.[11]

This balance sheet would look different if one were to take into account not only output, but also labor input, which would require a conversion into monetary value. "It is certainly our experience," wrote a member of the clergy in 1789, "that the total value for us from one cow over the course of a year is a matter of what that cow produces during the summer, since whatever the cow produces during the winter is counterbalanced by the hay that it consumes, if one also takes into account the effort of taking care of it and providing forage." According to the cleric's rough calculation, a cow consumed 3 to 4 *klafter* of hay (worth between 27 and 36 florins) in the stable over the course of the winter. He further calculated that the amount of butter produced during the same period was worth about 30 florins and that the cheese produced more or less compensated for the labor involved in foraging, so that the winter costs and benefits cancelled each other out. During the summer in the Alpine pasture, dairy products worth 11 or 12 florins could be expected from one cow; once the salaries of the cheese-makers and cowherds were deducted, about 10 florins remained. Other balance sheets confirmed the rule according to which the wintertime utility of a cow was taken up by the costs of hay consumed, if converted into monetary values. In general, one can conclude that forage production for winter stabling required land that was valuable, since its availability was limited, and also a large labor investment. On the other hand, Alpine pastures were broad and extensively exploited, providing forage with a minimum investment of labor. In other words, alpiculture represented a form of exploitation that consumed space but saved labor, with a favorable proportion of costs and benefits. Still, when one considers the length of the exploitation, the number of workers and the total product, the importance of alpiculture in the Grisons pales in comparison to the importance of valley-based agriculture, even if one only considers the livestock sector.

It is not easy to assess how widely the validity of regional findings such as these could be extended. One would need to examine many factors, some of which act in opposite directions. I will limit myself here to a few observations. If the parameters are the current sub-national political boundaries and the altitude of the settlements, the Grisons, whose communal centers are located at an average altitude of 1129 meters, is the highest region in the Alpine space.[12] Frequently large pastoral resources went hand in hand with high altitudes. In the Grisons at the turn of the last century, the percentage of pastures with respect to total surface area was estimated (according to different methods) at 41% or 55%. Another factor is that the average amount of time during which Alpine pastures were occupied by livestock tended to be significantly shorter as the altitude increased; thus, there were regions in which these durations surpassed those of the Grisons by 25% or more.[13] Still, such general conditions of alpiculture as these can only be partially explained by environmental factors, because the division between summer resources and winter resources was the result of a historical process that was neither inevitable nor uniform. One piece of evidence in this regard is provided by the frequent exceptions to the rule linking shorter pasture times to higher altitudes. Many Italian pastures, for example, were located at much lower altitudes than the ones in the Grisons, but animals were kept there for shorter periods of time. This is most likely related to higher demographic and agrarian density in the Italian areas.[14]

By partially anticipating some later arguments, we can take note of two things. First, between the sixteenth and nineteenth centuries, alpiculture was tied to consistent forage production and to winter stables in the valleys. It would be untenable to describe this kind of economic structure as archaic. If the goal was to minimize labor requirements, avoiding winter stabling wherever possible was necessary, and this was in fact done according to specific regional situations (see below). Second, it makes little historical sense to follow Malthus or Frödin in representing Alpine agriculture as a closed system that was determined by its most extensive element. In reality, processes of intensification could be established at lower altitudes and in turn exert notable influence on the exploitation of higher-altitude pasturelands. What characterized Alpine agriculture was not so much alpiculture *per se*, but the sharp intensity differential (economic exploitation that was intensive in some areas and extensive in adjacent ones).[15]

Chapter Three

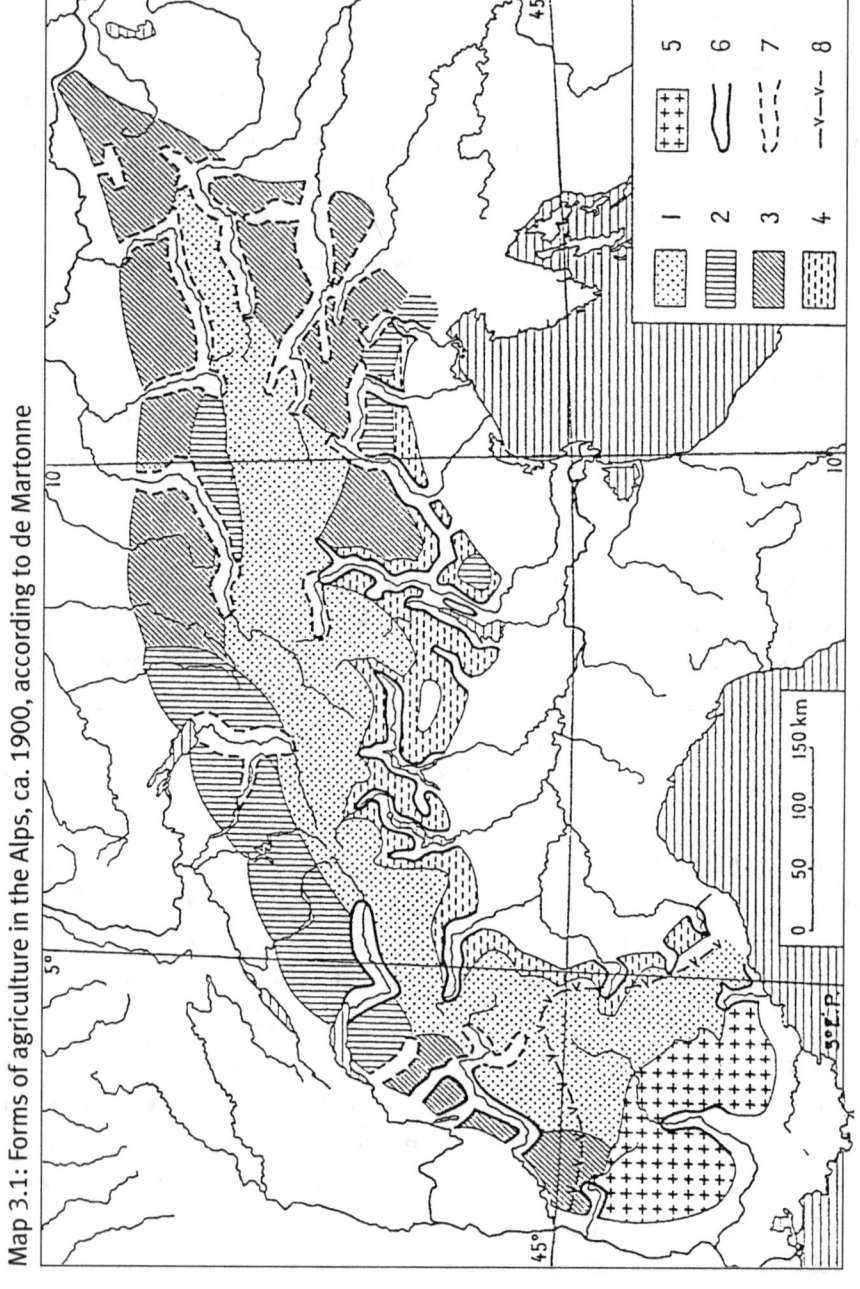

Map 3.1: Forms of agriculture in the Alps, ca. 1900, according to de Martonne

Source: Emmanuel de Martonne, *Les Alpes. Géographie Générale* (Paris, 1926), 157. Explanation of the legend in the text.

Agriculture and Alpiculture

Arguing from this perspective, the French geographer Emmanuel de Martonne—in a compendium that has now become a classic—identified a regional agrarian typology. Though his model is quite simplified and employs varied criteria, it still offers a useful panoramic view of the Alpine space in the period around 1900.[16] His explanations can be summarized as follows (see map 3.1, with its legend numbers):

> *Internal Alpine regions*: mainly large livestock-raising, grassland farming, also cereals, potatoes (1).
> *Sheep-raising regions*: sheep predominate south of line (8), partly kept in transhumant operation; very pronounced in the *pays à moutons* (5) alongside cereal farming.
> *Livestock-raising regions*: specialization in raising large animals, grassland farming (2).
> *Forest economy regions*: extensive woodlands with notable importance for the rural economy (3).
> *Insubrian regions*: mixed economy with chestnut, wine, and corn production; higher population density (4).
> *'Golfes de plaine'* ('flatland bays'): marked agricultural variety with corn and other crops in low-lying river valleys; dense population; viticulture is important in some regions (6), but of secondary importance or non-existent in others (7).

It is obvious that de Martonne's synthetic description and his map of *genres de vie agricoles* (forms of agrarian life) are reductive and only permit one to derive an initial feel for the problem. What emerges clearly is the wide variety of agricultural intensity to be found in the Alpine space at the end of our study period. How might it be possible to understand this variation in greater detail, and to think about it historically?

Cropping frequency and yields

The most important limiting factor for mountain agriculture was the growing season, which grew shorter as altitude increased. Steep slopes in many areas made agricultural work exceedingly difficult, but when compared to the state of the terrain in hilly areas near the Alps, and taking into account the technology of the time, steepness was less of a factor for the Alpine space than was altitude. The shortness of the growing season exerted an influence in various ways. According to agronomists' estimates, for example, annual grass production diminished by about 40% for every 1000 meters of elevation increase, and on farming plots the possibility of planting a second crop after the first was harvested was reduced. The longer duration of the cold season implied, conditions being equal, a longer duration of snow cover, making stabling a more pressing and expensive need.[17]

Modern histories of climate have demonstrated how much the climatic situation changed during the centuries of our study. Awful environmental conditions were frequent, especially between 1565 and 1629, and in the years around 1690. Bad weather did not seem to play as significant a role during other periods, and in the late nineteenth century there was a warming of the climate as documented by evidence of glacial shrinking.[18] While the climatological approach is of great help in explaining production variations over the short term, its importance for agrarian developments over the long term is controversial. Given the variety and diversity of unfavorable factors and of agricultural forms, it is not quite clear whether a region as vast as the Alpine space should be characterized as generally any more crisis-prone than surrounding regions. From a methodological perspective, one should keep in mind that the sensitivity to climate at the high-altitude limits of certain crops or human settlements—a term dear to traditional geography—is of restricted importance. Only a very small part of the Alpine population lived in such areas.

A useful criterion for measuring agricultural potential, or actual agricultural production, is provided by the concept of cropping frequency. This refers to the number of times a given piece of land (not necessarily one defined as "cultivated land") was exploited within a given period, with reference either to field cultivation or to hay meadows.[19] A whole series of topographical descriptions and other sources describe the differing productivities of forage-producing fields in the Alpine space. Let us follow Johann Jakob Staffler, who in 1839 published a meticulous account of the Tyrol (as it existed under the Hapsburgs) and the Vorarlberg. According to him, three categories of forage-producing fields could normally be identified there: precocious, late, and Alpine ones. "Precocious meadows in valley bottoms provide an abundant harvest of grasses. In the valley of the Inn, the Wipptal, Val Pusteria, and Val Venosta these are cut two times, and in some of the best areas, three times. . . . At Merano, Lana, and Bolzano, where the fields are worked very carefully, there are on the best lands four harvests—*Heu, Grumet, Povel,* and *Nachpovel*—that are said to be so large that from one *Jauch* (about 1.4 acres) of land, 80 to 100 *Centner* of forage is produced." Unlike precocious fields, late fields were not fertilized and were only cut once per year. "Finally, Alpine fields, which are neither fertilized nor irrigated and are simply exposed to nature, are not always cut each year, but in many places are only cut every other or every third year. The hay from this mowing is very thin and short, often barely a span long, and is usually harvested at the rate of only 2–6 *Centner* per *Jauch*." Since some areas "contain enormous expanses of these Alpine fields," the total produced was not negligible.[20]

It is possible to collect examples of hay production in high-altitude and steep lands for many parts of the Alpine space. Production was subject to sharp oscillations and was sometimes carried out only sporadically. This harvest rhythm, from slow to intermittent, contrasted starkly with the economic systems of the plain of the Po, where various forage crops were harvested up to seven or eight times per year during the early modern period.[21] There were also consistent differences in cropping frequency within the Alpine space itself. The extreme figures indicated by Staffler vary by a factor of twelve. It is plausible to assume, and it can be demonstrated by some available records, that these differences in field use were already widespread at the beginning of the early modern period. The dimensions of these differences increased over time, however. In any case, we are continually confronted with cases where, between the sixteenth and nineteenth centuries, harvest frequency increased, especially on valley bottoms.[22]

Grass-growing and hay-cutting possibilities were influenced by altitude and by other climatic factors, but also in considerable measure by cultivation practices themselves, including fertilization and above all irrigation. Available descriptions show that there were noteworthy irrigation techniques in many places, even in wetter areas on the northern slope of the Alps, where irrigation was designed to enrich and warm up the land early in the growing season. With respect to the intensity of irrigation, there were naturally great regional and temporal differences. The amount of labor necessary to build the infrastructure is an indicator that happens to be historically useful since it is easily documented. Notable examples include the 500-meter long tunnel at Exilles (Piedmont), constructed between 1526 and 1533 to redirect the course of a mountain stream, and the 600-meter long aqueduct at Laas/Lasa (Südtirol/Alto Adige), constructed in 1790 to carry water out of the valley. On the basis of a list of important irrigation canals, with their approximate or exact dates of construction, one can roughly identify the phases of the historical economic conjunctures for the Valais. Of the canals for which records exist, a great many were built during the fifteenth and sixteenth centuries (21), very few were constructed during the eighteenth century (4), while during the nineteenth century there was again frenetic construction activity (21).

As far as we are aware, this chronology corresponds to the demographic trends of the region. In other areas as well, the correlation between population levels and agrarian intensification in the form of irrigation activity seems to be particularly close.[23]

Part of the explanation for these phenomena is to be found in an analysis both of the labor burden being sustained and of outputs. Sebastian Münster's *Cosmographia* tells us that the Valaisans were already saying in the sixteenth century that "water costs them more in money and labor than does wine." One study from the early twentieth century gives the example of a farmer who owned four large animals and had to irrigate 160 times—work that required 80 days of labor. Thus, the labor needed for irrigation was at least double that needed for hay-making, but forage production stood to increase remarkably as a result. Existing studies tell us that irrigated fields, under certain conditions, produced five or even eight times as much as non-irrigated ones.[24]

Cropping frequency for farmland other than hay fields also varied significantly in the Alpine space. In Styria and in lands surrounding the eastern Alps, the so-called *Brandwirtschaft* (a kind of swidden or shifting cultivation) was among the oft-employed techniques during the period studied here. In the 1830s and 1840s, it was estimated that about a quarter of the grain produced in Styria was grown with this method. This economic form was concentrated in the mountainous regions of Upper Styria, especially in areas with isolated farms and extensive woodlands. Inhabitants of these forest areas engaged in the intensive cultivation of fields near their farms, and each year, according to their resources and needs, they also created new fields in places that were exploited on a cycle of fifteen (or up to fifty) years. First they felled the trees that had grown in the meantime and used the best lumber for their own needs or for commerce, and then they incinerated the remaining growth in a controlled burn. For a brief period the burned field was used for cereal production, and then as pastureland, and finally it was allowed to re-forest. Historical studies show how during the sixteenth and seventeenth centuries this and other methods spread widely and then diminished in the eighteenth century under the regulation of new state-sponsored forest legislation. In Styria, this kind of farming was important enough to be included in legislation, and it survived until after 1950. In very different places and times, one finds similar techniques. For example, they were important in places on the edge of the French Alps, whose rich woodlands were identified by de Martonne (see map 3.1, zone 3). The western part of the Vercors, for example, was for a long time "a land of *essarts*, that is, of farming on burnt land, practiced throughout the forests."[25]

Agriculture and Alpiculture

Apart from burnt-field methods, and with strict reference to permanent crop cultivation, harvest frequency in the Alpine space varied by a factor of about four. Well-documented and frequently studied are those systems in the western Alps, and in particular in Provence, that were characterized by leaving the fields fallow every second or third year. References to these systems, combined in various ways and alongside other forms of cultivation, begin in the late Middle Ages. The fact that this rhythm underwent only minimal variations during the eighteenth and nineteenth centuries is typical of the *Alpes du sud*, especially when compared to the neighboring mountainous regions of Savoy-Piedmont, which during the same period shifted to more intensive exploitation methods. The maintenance of fallow practices was linked to specific demographic situations. In 1801, despite its relatively average altitude, the department of the Alpes-de-Haute-Provence counted 19 people/km^2 and in 1901 only 17 people/km^2, while the departments of Savoie and Haute-Savoie had densities of 35–40 (1801) and 41–57 (1901) people/km^2.[26]

Let us consider the example of shorter harvest sequences from some regions in the central Alps. In the Engadine, from the beginning of the early modern period if not earlier, the fields were sowed annually. Only exceptionally can we find records from this high valley of years when fields were left fallow—for example, when there were fertilizer shortages. Often, during the Old Regime, observers from the north took note of "double harvests" in lower-altitude areas on the southern slope of the Alps. In 1706 Johann Jakob Scheuchzer remarked enthusiastically that the Valtellina participated "in the excellent amenity and fertility of Italy." Another researcher from Zurich reported in 1786 that in the lower Ticino grain was grown everywhere: "no, almost everywhere the same field is sown twice, but not with the same grain, or even, besides grain, things that are very productive, like corn, foxtail millet, millet, and so on."[27]

The history of multi-cropping can best be followed by looking at the diffusion of new plants, which were often used in second sowings: buckwheat from the fifteenth century, corn from the late sixteenth (see below). Among the important peculiarities of the second crops was their relatively brief vegetative period. What made similar increases in harvest frequency more difficult and ultimately impossible in higher-altitude areas was for one thing the smaller number of cultivable plants and a slightly longer vegetative period, and for another thing the shorter growing season due to environmental conditions. It is at present difficult to establish precisely how new plants were excluded from a given elevation level. One geographic study looked at this issue in the highest-altitude areas, and identified a specific altitudinal fallow. The thesis is based on the

Chapter Three

observation that in the high valleys of the western Alps winter grains were often sowed every other year, and that the grain remained in the fields too long for annual sowing. Without denying the unfavorable weather conditions, I consider this interpretation to be exaggerated. Even the study's author admits that plants with shorter vegetative periods were known in this area, as were techniques for accelerating the growth cycle. The study also refers to heavy labor demands during certain seasons, but underestimates the influence of demographic factors on the process of intensification. It would be important to interpret fallowing practices in the total historical context of the western Alps, in which there were a number of similar cultivation methods.[28]

And what about yield sizes? It is not easy to provide a response. There is a tendency to assume that output was small, but there are no comparative studies and the fragmentary data do not provide a clear picture. Statements from various parts of the Alpine space during the early modern period concerning the ratio between seeds planted and harvested do not show systematic differences between higher-altitude lands and valley bottoms or contiguous regions. Sometimes, for example, valley wheat production was higher while the production of rye was greater in mountainous areas. Similar impressions are provided by more systematic and somewhat more reliable nineteenth-century estimates, which referred to output per hectare. According to data collected in 1847, in the mountainous areas of the canton of Berne the output in terms of a given surface area for various types of grain was not inferior to that of the *Mittelland* (high Swiss plain). In the Alpine territories of the Hapsburg crown average yields between 1789 and 1880 appear to have been slightly larger than in the non-Alpine regions of Lower and Upper Austria, while after this period the relationship reversed itself. General economic factors could in any case have a strong influence on harvests. Fertilization was important, for example, and this was an area in which mountain zones often had an advantage from the start.[29]

The range of harvest frequencies—by way of summary—both for grass-growing fields and for cereal production, varied from sporadic exploitation to multi-cropping: it was thus very broad. In zones of intensive farming, grass and grain production were combined in a variety of ways.[30] In extensively-used lands, the two crop types blended with pastures and woodlands; this was particularly fluid when it came to Alpine forage production and swidden cultivation. Given this situation, it would be a mistake to give too much importance to the difference between cultivated and non-cultivated land, or between agrarian intensification and agrarian expansion. Apart from mountain peaks, which began to be of interest for climbers only in the late eighteenth and nineteenth centuries, from the beginning of the early modern period onward there were probably very few territories that were not subject to exploitation in some way or another. A close look at reports concerning the use of 'new' lands shows in fact that in most cases these lands were undergoing a transition from extensive to intensive use.

Since most of these reports refer to wide river valleys at modest altitudes, one may suppose that this zone enjoyed above-average agricultural growth. The topographies of these bottomlands could change from year to year, and were usually farmed very extensively. When local populations initiated and carried out the construction of dikes and small canals, there were in some cases significant increases in agricultural output during the early modern period. The eighteenth and even more the nineteenth century saw the beginning of a period in which river courses were changed and large-scale land improvement projects were undertaken. These initiatives became programmatic in nature and were more and more frequently supported and decided upon by state authorities. The specific circumstances in which individual bottomlands were transformed varied from case to case, but before 1900 this affected all of the large river tracts and the *golfes de plaine*: Durance, Isère, Rhône, Rhine, Inn, Salzach, Enns, Mur, Drava, Tagliamento, Piave, Adige, Adda, Ticino, Dora Baltea, and many others. Favorable climatic conditions benefited agriculture in these regions. One should note, however, that it was necessary not only to construct the necessary infrastructures, but also to maintain them.[31]

The intensification of animal husbandry...

The process of agrarian intensification is amply documented by changes in the array of Alpine livestock and domesticated plants. Let us leave aside for the moment the issue of how the two sectors relate to each other and instead consider each area separately.

A number of recent studies focusing on the mountainous areas of Switzerland have analyzed, in a variety of ways, the quantitative shift away from sheep-raising to cattle-raising—an important phenomenon that began to be apparent during the Middle Ages. Nicolas Morard and Hans Conrad Peyer, working within the context of the history of the pre-Alps of Fribourg, have argued that cattle-raising became predominant in the northern Alpine regions at the beginning of the early modern period, but that sheep-raising remained important especially in high-altitude areas. According to a rough estimate, in the seventeenth century there were 600,000 Swiss sheep, then 500,000 in the early nineteenth century, and only 220,000 in about 1900 (of which

Chapter Three

two thirds were located in mountainous cantons). A competing interpretation holds that cattle were important at least as early as the late medieval period, minimizing the importance of the shift from sheep- to cattle-raising and locating it during these earlier centuries. Opinions seem to depend partly on precisely which geographic area is being investigated, which suggests that in reality this shift took place in different ways from one area to another. On a regional scale, it is possible that there really were revolutionary developments within the composition of animal species before and after 1500, but if a single state, or even more, the entire Alpine space is taken into account, the process must be considered a long-term tendency.[32]

Examples of this tendency may be taken from many places for the sixteenth through the nineteenth centuries. In Slovenia, data from the years 1510 and 1630 exhibit a sharp reduction in the number of small livestock and a marked increase in cattle. Thereafter, this change is visible on a much wider scale: from 1630 until the end of the nineteenth century the number of ovines seems to have decreased by about 50% in the mountainous area, while the number of bovines continued to grow. In the area near Belluno, which belonged to the Republic of Venice, the number of small livestock only decreased between 1564 and 1780 from 41,000 to 37,000 (probably with a shift from sheep to goats, which at the later date constituted a third of the total); large livestock on the other hand registered a clear increase, from 14,000 to 20,000. As in large portions of the Italian Alps, in the nineteenth century this trend accelerated. From Savoy we find an abundance of numerical indications for the early modern period. In the Maurienne, an average household had fewer small livestock in 1759 than in 1561, while over the course of the same period they had a third more bovines. This shift also occurred in the Beaufortin to as radical a degree as it had in Fribourg, though starting at a later point. Of the animals stabled in the winter of 1561, 46% were sheep and 18% were milk cows. In 1758 the percentages changed to 9% and 38% respectively.[33]

These last figures point to the decisive upturn in cheese-making in the Beaufortin, which from about 1600 had increasingly specialized in the market-oriented production of hard cheeses. Among the different forms of the intensification of animal husbandry, this was an obvious case, but the growing importance of cattle-raising was also linked in other ways to increases in production levels and labor demands. The latter resulted mainly from the winter stabling of cattle, which was more intensive than for sheep because it lasted longer and required more forage.

On the regional level there was one significant exception to the broad shift described here: in the southern French Alps, and in Provence in particular, sheep-raising remained extremely important as late as 1900 (see map 3.1, zones 5 and 8). Available data suggest significant continuity here over the long term. In some areas of Provence, both in the late fifteenth century and in the late nineteenth, cattle were only a very small part of all livestock. Sheep-raising in southern France is most often discussed in the context of transhumance, which was based on a combination of winter pasturing in the plains and summer pasturing in the mountains, making it necessary to move animals long distances. Geographic studies have often considered transhumance to be a paradigmatic case of a primitive kind of economy whose form was determined primarily by environmental conditions. Historical research has shown, however, that these migrations began in the later Middle Ages and originated in mountain regions, which sought additional sites for wintering livestock. Then, during the fourteenth and fifteenth centuries, the opposite movement from the plains to the summer pastures became more and more frequent. The development of transhumance has been related to the demographic collapse during this period. Population loss is said to have favored the emergence of unused pasture areas in mountain regions in conjunction with the simultaneous increase in meat consumption, especially in the plains.[34]

Demographic factors can also be cited to explain how the system was maintained. As mentioned above, in the southern French Alps population density was quite low even when it reached its maximum level in the early nineteenth century. Equally important must have been the existence of unused pasture areas in the low plains, especially in the area around the mouth of the Rhône. In northern Italy, in any case, the reduction in amounts of available pasture land due to agrarian intensification was a decisive reason for a decline in sheep transhumance, which had been widespread between the plains and the Alps. Political evidence of this development included the 1765 decree by Venetian authorities that regulated and limited pasturing rights in the entire Terraferma.[35] Along with the decrease in pasture area, from the sixteenth century onward there emerged new and more intensive forms of transhumance. These were undertaken by an entrepreneur known in Italy as the *malgaro* or *malghese*, and north of the Alps as *Küher* (cow-men), pointing to the fact that cattle had now become centrally important. Unlike previous practices tied to sheep transhumance, in this new system the transhumant cattle were kept for a certain length of time in stalls and fed there. In places where, in the eighteenth century, permanent stabling began to emerge, the basis for this kind of livestock movement disappeared also.[36]

It is difficult to find evidence explaining the change in the numbers of goats among Alpine livestock, since references are often only negative ones. Goats were held to be productive, low-maintenance animals that provided more milk than cows (given the same amount of forage), and that required less land. "Since they are not expensive to acquire, and since they can consume a greater variety of lower-quality forage, goats can be maintained with fewer expenses and work," wrote Staffler about the old Tyrol, "so that it is mostly poorer families that have a few goats, either without or with a single cow." Goat-raising methods depended on whatever kinds of collective resources were available, and, in conjunction with the specific social-status configurations involved, often gave rise to conflicts, limitations, and prohibitions. An edict issued by the *Parlement* of Dauphiné in 1565 prohibited goat-raising in the lowlands. Between 1672 and 1698 the court published three decrees extending this prohibition to cultivated areas in the mountains. Only rarely-used and hard-to-reach areas were to remain open to goat pasturing. Litigation records and other documents make it clear, though, that these edicts were ineffective—indeed, they should probably be interpreted as proof of the intensification of goat-raising. In the Beaufortin, the number of goats as a percentage of all livestock increased in parallel fashion to the increase in cattle and the decrease in sheep. In 1561 16% of all stabled livestock were goats, while two centuries later goats represented 35%. Analogous processes presumably took place in other areas, though such developments were always strictly dependent on the social context. If one is to believe sporadic data, the great era of goat-raising was the period from the late eighteenth through the early nineteenth century.[37]

... and of plant cultivation

The cultivation of comestible plants in the Alpine space shifted during the period considered here due to the expansion or decline of old crops (like chestnuts, grapevines, and fruit), and to the introduction of new ones (like corn and potatoes).

Chestnut-growing, which had been documented in antiquity, was carried out mainly on the southern slope of the Alps, especially in western regions. Throughout the Middle Ages it seems to have developed significantly, a process that continued and perhaps even accelerated after 1500. Piedmontese cadastral records from 1699 and 1752 indicate that the expansion of chestnut cultivation there continued through the beginning of the eighteenth century. At that time, in some valleys in northern Piedmont, chestnut trees were by far the most important plants. Among their benefits were high yields per surface area and the fact that they could be planted, by means of terracing, on the steepest slopes. During the eighteenth century, chestnut cultivation in southern Piedmont began to give way to vine-growing, while in the previously depopulated Waldensian valleys it seems to have accompanied the new waves of settlement.[38] Buckwheat, a polygonum that works well in second harvests because of its short growing cycle, is another plant whose Alpine career does not follow the conventional periodization. It seems that southern Alpine areas (not only in Italy) became familiar with buckwheat during the fifteenth century. In Slovenia its cultivation was mentioned in 1426 and it appears to have been widely diffused by 1500. We find buckwheat in Carinthia in 1442, but some areas cultivated it much later—at the end of the seventeenth century in the Vorarlberg, for example.[39]

The earliest references to corn (maize) cultivation in the Alpine space also come from the southeastern areas. In 1572, in the parish of Thal in Styria, west of Graz, there was a conflict over the poles used by the peasants for growing corn. The poles had been illegally taken from woods belonging to a lord. The regulations of the Graz millers from 1608 indicate that the new cereal soon became important in the area. In the district of Salorno, located at what is now the boundary between the Trentino and the Südtirol/Alto Adige, *türggischer Weizen* (Turkish grain, or corn) appeared in tithe accounts from 1573. Around 1600, the cultivation of corn was confirmed by a topographic description of the same jurisdiction explaining that this cereal was consistently produced all over the district and it was "good for cooking and for bread-making." Supply lists from the northern Tyrol dated 1615 show that corn appeared in some communes there (whereas earlier, it was barely ever mentioned). At around the same time it was introduced to the area near Belluno, where an agronomic text of 1633 praised this plant of American origins as "the treasure of our country," which helped the poor survive, gave middle-income people new strength, and "filled the money purses" of the wealthy. In that region "everyone is involved in such work [farming this plant]," to the advantage of the Bellunese. Later more critical opinions began to be voiced. In 1797 a tract argued that corn took up too much farmland in the area near Belluno—an indication that this crop had now become very important indeed. In fact, many sources show that the rapid expansion of corn cultivation took place not during the initial phase, but during the eighteenth and nineteenth centuries, especially between 1750 and 1850. In the Tyrol and the Vorarlberg, lands of the Austrian crown, cultivation appears to have increased by almost 50% between 1809 and 1839. During this period, in the two districts of the present-day Trentino, corn accounted for over 25% of all comestible field crops. From the perspective of the surface area occupied, corn was, at least at the end of our study period, the region's predominant plant.[40]

Chapter Three

Studies of this crop have been sporadic and often imprecise, making it difficult to provide a coherent chronology of corn-farming in the Alpine space; the same can be said for potato-farming. Table 3.1 includes all of the regions, according to present-day administrative boundaries, 75% of whose territories are located inside of the Alps. The table indicates the date at which these two plants began to be cultivated on a considerable scale in the given region. Despite its 25-year periodization scheme, the description is approximate, as can be deduced from the uncertain dates listed in the records. The table makes it clear that over the course of its long history corn was of only marginal importance (sowed in small plots or statistically irrelevant) in several regions until the end of our period. The introduction of the potato proceeded differently. In the sixteenth century it was known among Europe's scholarly circles, and was planted in most Alpine regions from the second half of the eighteenth century on, becoming widely diffused in the entire Alpine space by the nineteenth century. Whereas corn could only be grown at relatively low altitudes, especially during the early modern period, potatoes could be cultivated in true mountain zones.

The fact that potatoes were introduced in some of these zones quite early on and became the dominant crop there prompted some authors to conclude that potato cultivation was generally more important in the mountains than in surrounding areas. Moreover, Pier Paolo Viazzo believes that the potato spread to the Alpine regions of Austria later and to a lesser degree than in Switzerland, owing to differences in social structure.[41] The quantitative data required for pursuing these two questions were first recorded at different times in various countries and regions. The first country-wide collection of farming data in Switzerland was not carried out until 1917, making it difficult to test Viazzo's thesis. Still, the sources show that in the nineteenth century per capita potato production among the rural population was actually higher in some areas surrounding the Alps than in neighboring mountain zones. With respect to the Austrian Alps, a particularly late introduction of the potato in Carinthia might be seen as evidence of Viazzo's argument. But this region had already introduced clover in about 1700, apparently steering itself in a different direction toward agricultural intensification. In the years around 1910, Carinthia counted 4.2 *Ar* (1 *Ar* = .0241 acres) of potatoes for each member of the rural population. In the Swiss Alpine cantons the maximum cultivation in the period soon thereafter was 3.8, and the average was only 2.5. Thus, neither of the two hypotheses mentioned above can be confirmed by these indications.[42]

Agriculture and Alpiculture

Table 3.1: Introduction of corn and potatoes in the Alpine regions, 1500–1900

Region	1500	1600	1700	1800	1900
1 Alpes-Maritimes (F)				€P ═══	
2 Alpes-de-H.-Provence (F)				P────	
3 Hautes-Alpes (F)				P────	
4 Savoie (F)				CP ═══	
5 Haute-Savoie (F)				CP ═══	
6 Imperia (I)				€P ═══	
7 Val d'Aosta (I)				€P ═══	
8 Sondrio (I)				€ ── P ═══	
9 Trent (I)		€ ─────────── P ═══			
10 Bolzano-Bozen (I)		C ─────────── P ═══			
11 Belluno (I)			C ───────────────		P ═══
12 Valais (CH)				€P ═══	
13 Ticino (CH)			C ── ──── P ═══		
14 Grisons (CH)				C P ═══	
15 Uri (CH)				P────	
16 Unterwalden (CH)				P────	
17 Schwyz (CH)				P──── €	
18 Glarus (CH)				P────	
19 Appenzell (CH)				P────	
20 St. Gall (CH)			C ─── ──── P ═══		
21 Liechtenstein (FL)			C ─── ──── P ═══		
22 Vorarlberg (A)			C ── P ═══		
23 Tyrol (A)			C ─── ──── P ═══		
24 Salzburg (A)				P	
25 Carinthia (A)			C ────────	P ═══	
26 Styria (A)		C ─────────── ──── P ═══			

C, P = Corn, potatoes under significant cultivation before 1900. €, P̶ = questionable.

Sources, cited in abbreviated form: (1–3) F. E. Foderé, *Voyage aux Alpes Maritimes*, (Paris, 1821), 2: 43; *Le monde alpin et rhodanien* 1987, 117–133; Blanchard 1938–1956, 5: 423. (4–5) Nicolas 1978, 2: 691–694. (6) see (1) and G. Felloni, *Popolazione e sviluppo economico della Liguria nel secolo XIX* (Turin, 1961), 15. (7) *Bulletin de la Société de la Flore Valdotaine* 7 (1911): 98–100; L. Jacquemod, *Trefolle, tartifle, pommes de terre en Vallée d'Aoste* (Aoste, 1993), 53–67. (8) P. Ligari, *Ragionamenti d'agricoltura* (Sondrio, 1988), 72–77; Biblioteca Ambrosiana di Milano, L 42 inf/7, f. 26. (9–10) H. Telbis, *Zur Geographie des Getreidebaues in Nordtirol* (Innsbruck, 1948), 30, 33; H. Penz, *Das Trentino* (Innsbruck, 1984), 197; *Berichte zur dt. Landeskunde* 12 (1954): 217–218. (11) L. Messedaglia, *Il mais e la vita rurale italiana* (Piacenza, 1927), 150; A. M. Bazolle, *Il possidente bellunese* (Feltre, 1987), 2: 72. (12) Archives d'Etat du Valais, Fonds de Courten, Livres 14; Netting 1981, 162. (13) Guzzi 1990, 126; Archivio Storico Ticinese 111 (1992): 76.

Chapter Three

From an overall perspective of the entire study period, the differences between specific areas are certainly less important than the general difference between the availability of botanical resources in the sixteenth century and their actual use during the eighteenth, and even more so the nineteenth, centuries—a difference that applies to both corn and potatoes. It is noteworthy that their yields are far superior to those of traditional cereals, though harvests could vary significantly. Substantial increases in output were made possible above all by massive increases in labor inputs. In demographic terms, it can be argued that only the growth of the population during the eighteenth century made this development rational—that is, both necessary and possible.[43]

With viticulture and fruit-growing, intensification was closely tied to large-scale integrative processes. In many northern valleys viticulture contracted while fruit-growing expanded; in the south, on the other hand, both forms continued to expand, with the exportation of southern wine northwards growing in importance. There are references to such developments for many parts of the Alpine arc, but with notable differences of chronology and scale. Here we will limit ourselves to the middle section of the Alps.

As indicated in map 3.1, viticulture in the Rhine and Inn valleys was inexistent or only marginally important in around 1900, unlike the situation in the *golfes de plaine* south of the Alpine watershed (zones 6 and 7). This arrangement must have come into being for the most part during the early modern period. The Valtellina was among the older wine-producing regions, but the amount of its territory covered by vineyards expanded rapidly from the second half of the seventeenth century onward, requiring new terracing reinforced by walls. Soon, sunny hillsides there were for long stretches "covered with expertly worked vines, supported by walled terraces," as described in a text of 1716. Despite incipient criticisms about being forced into viticulture, the increase continued, even after possession of the Valtellina changed hands in 1797.

---

(14) J. A. v. Sprecher, *Kulturgeschichte der Drei Bünde im 18. Jh.* (Chur, 1976), 77; *Bündner Monatsblatt* 1982, 117–47. (15) J. Bielmann, *Die Lebensverhältnisse im Urnerland während des 18. und zu Beginn des 19. Jh.* (Basel, 1972), 179. (16) S. Bucher, *Bevölkerung und Wirtschaft des Amtes Entlebuch im 18. Jh.* (Lucerne, 1974), 168. (17) Bircher 1938, 13. (18) F. Kundert, *Die Lebensmittelversorgung des Landes Glarus bis 1798* (Glarus, 1936), 93. (19) Hp. Ruesch, *Lebensverhältnisse in einem frühen schweizerischen Industriegebiet* (Basel, 1979), 1: 112. (20) *Montfort* 3 (1948): 83; E. Gmür, *Die Geschichte der Landwirtschaft im Kanton St. Gallen* (St. Gall, 1907), 13. (21) Ospelt 1972, 164–65. (22–26) *Sozialer und kultureller Wandel in der ländlichen Welt des 18. Jh.*, ed. E. Hinrichs (Wolfenbüttel, 1982), 166–77; R. Sandgruber, *Österreichische Wirtschaftsgeschichte vom Mittelalter bis zur Gegenwart* (Vienna, 1995), 158–59; *Carinthia* 177 (1987): 243.

In the Trentino it appears as though surface area occupied by vineyards quadrupled between about 1500 and 1700, and this tendency continued during the eighteenth and nineteenth centuries. Along with the Südtirol/Alto Adige, this region now produced almost all of the wine in the Tyrol, because in the northern Tyrolese valley of the Inn, until the eighteenth century, viticulture disappeared under the pressure of wine imports. Similar consequences, though perhaps less stark ones, were created through the competition of Valtellina wine in the Vorarlberg and in other parts of the Rhine valley. In both areas, though, fruit-growing expanded, beginning during the Old Regime and continuing until the present.[44]

After this overview of developments in livestock-raising and farming, it now remains for us to consider the two sectors together. From what has already been presented, one can imagine that the relationship between the two could vary notably from region to region. Until the late-nineteenth-century transportation and market revolutions it is difficult to identify general trends valid for the entire Alpine space. For some regions it would not be a mistake to posit a long-term process of agrarianization. The opposite development has been better studied, especially in a north Alpine zone running from Savoy to Austria, via Switzerland (see map 3.1, zone 2). From the late Middle Ages onward, in this *Hirtenland* (pastoral region)—as it came to be called in Germanophone literature, locally at first, and then more widely—grain-growing became secondary to animal husbandry, which sometimes completely replaced it. A notable example are the pre-Alps of Fribourg at the start of the early modern period, where grain fields began to vanish from the landscape bit by bit and flocks of sheep thinned out as farmers specialized in cattle-raising. Several authors see in this specialization an example of extensification, but without offering any proof: they simply observe that cattle-raising is less labor-intensive than cultivating grain.[45]

One could object to this claim as follows: generally grain-farming and animal husbandry could be practiced with very different intensities—the history of the Alpine space amply illustrates this fact—and the latter was not necessarily always more extensive than the former. Indeed, in the particular case of the displacement of grain-farming by livestock-raising, as we have seen in Fribourg, I hold the claim to be unrealistic. In the earlier period, space-consuming plant cultivation was carried out alongside significant pastoral activities. The increase in cattle-raising that followed required intensive grass-growing, which in turn involved fertilizing and other preliminary activities, as well as two to three cuttings per year. Stable-feeding also required substantially increased amounts of labor; and to this one must add the work involved in hard cheese production. Later, potato cultivation also began. And how did the population develop? According to available figures, it increased by a factor of 3.5 between roughly 1500 and 1811 over the territory of Fribourg (the detailed internal demographic developments are unknown). Nineteenth-century figures show us though that the population of the pre-Alpine districts doubled.[46]

Many facts suggest therefore that specializing in large livestock-raising increased rather than decreased the regional volume of labor. Even the formation of the so-called *Hirtenland* could in other words be considered a form of agrarian intensification (something that this Enlightenment-Romantic definition, which only took the limited pastoral component into account, dismissed from the start). The long-term growth in external demand must be considered as a crucial factor in this process of intensification and specialization. Generally, the Alpine animal stock was linked to markets on a much wider scale. Animals and animal products were without a doubt the most important goods that the Alpine arc placed on the market during the period studied here—we will return to this theme in chapter 5 in a context to be described below. It remains for us now to return our discussion of the technical dimensions of these agricultural developments.

Technology

Two questions are centrally important here: to what degree was Alpine-specific agrarian technology employed (such as technology adapted to very steep plots of land)? And what role did this technology play in the general history of Alpine agriculture? Examples related to these questions will be taken from a wide range of peasant methods and tools, and as in the previous sections of this chapter, they relate first to animal husbandry and then to plant cultivation.

If agricultural transport is understood in the widest possible sense to include the spatial relocation of goods, animals, and persons within a farm, then Alpine agricultural transport frequently required an above-average amount of labor. For one thing, the surface area worked by the farm, when involving a certain amount of livestock-raising and the exploitation of extensive high-altitude resources, was spread out over very large swaths of land. For another thing, the differential vertical distribution of land that was frequently hard-to-reach required considerable amounts of extra work from the persons concerned. In some areas it has been calculated that over half of all agricultural labor time was used for transport. One way to assess the historical

Agriculture and Alpiculture

importance of this kind of transport is to investigate the property inventories that list and assign value to means of transport or to rural construction.[47]

These two categories were functionally related to each other, and together formed the material portion of the agricultural organization of space. Differing investments in transport systems and rural buildings in the area under investigation permit us to distinguish between several paths of infrastructural development. We will use as a criterion choices related to the provision of winter forage, that is, the various activities connected to hay—the most important rural product in terms of volume.

(1) Investment in transportation: construction and maintenance of roads; vehicles and draft animals for transport over long distances; immediately after being harvested hay is transported to a central location, where it will later be used for feed.

(2) Investment in buildings: construction and maintenance of barns spread out across the territory, permitting decentralized storage and fodder provision; transport by foot (with back carriers) over short distances; the animals go to where the fodder is rather than vice versa.

(3) Intermediate form: temporary storage of fodder in the open or in decentralized barns; later transport and foraging in a central location; transport is merely delayed, not avoided.[48]

Past and recent studies on this theme have concentrated on the most visible systems, the decentralized ones (form 2, first phase of form 3), and have usually explained these systems in terms of the obstacles to vehicular transport presented by the Alpine landscape. According to a detailed description from the eighteenth century, in the Ticino, where there were a great number of isolated rural buildings, agricultural products were often transported with back carriers. The author noted that topographical conditions prevented vehicles from being used in many places. Only in some of the flatter areas of this southern Alpine territory could one find simple carts pulled by oxen. These forms of transport, which differed even across short distances, were similar to those documented in other areas and clearly lead back to the relationships between technical apparatus and terrain. One could also identify ways in which systems of intermediate open-air fodder storage, or of decentralized barns, were conditioned by the environment, especially when transport then had to proceed down long downhill paths. This was easier to effect by means of sliding techniques on winter snows, rather than immediately after

Chapter Three

haying on dry ground. Delaying transport in this way and using the so-called *Heuzug* (haysled) saved substantial amounts of labor in certain cases.[49]

Alongside indicators of the importance of natural conditions, one could oppose others that do not fit into the currently dominant interpretive model. Precisely this intermediate storage system was not motivated everywhere by the use of sliding transport techniques. In some regions it was adopted regardless of the topography and even at a relatively short distance from where forage would later be provided. Comparative observation shows generally that besides the micro-level differences there were macro-level ones, pointing to important historical correlations that have not yet been sufficiently investigated.[50] The most complete collection of materials related to rural transport was made by Paul Scheuermeier, who, immediately following the First World War, traveled through most of the Italian Alpine valleys and many of the southern Swiss ones, carrying out linguistic and ethnographic research. According to his observations, single sections of this widespread region were characterized by different dominant technologies. To simplify greatly, one could describe the postwar situation as follows: transport by pack animals in the west (Piedmont), human transport with back carriers in the central region (from the Val Sesia to the Val Camonica), and vehicular transport in the east (in part of the Grisons, in the Tyrol and in the Trentino).[51]

With respect to historical developments after the sixteenth century, one could point to evidence both of continuity and of change. The numerous farm inventories from one central region (Carinthia) show us, for example, that transport vehicles were a stable component of agrarian technology in early modern times. It is also interesting that the well-known eastern Alpine passes could support cart traffic from the beginning of the early modern period and even earlier, while the central and western Alpine passes were accessible to vehicles only after the construction of roads around the turn of the nineteenth century. But there were also important changes that took place over the course of this long period. During the eighteenth and nineteenth centuries in the western Alpine regions, trading with pack animals flourished, growing in popularity. In addition, cart traffic increased everywhere and, especially in the late nineteenth century, began to be used regularly even in Alpine valleys where previously such use would have been unimaginable.[52]

This triumph of vehicular traffic has created problems, in different ways, for retrospective interpretations of past forms of transport. Differently from what one might initially suppose, "archaic" transport by means of back-carriers was in fact an efficient technology in certain circumstances. This has been documented especially in those regions where contiguous territories had long confronted each other with completely different systems for organizing space. In the Grisons, agricultural writings show that carrying hay by back into barns that were spread out across the territory at short distances from the meadows was much more rapid than vehicular transport. Until the beginning of the nineteenth century it was far from certain in this region whether the centralized or the decentralized system (forms 1 and 2) would be preferred on the whole. For the most part, comparisons made by economists resulted in series of advantages and disadvantages that seemed to balance each other out depending on the specific case. In a way, though, their discussion was a theoretical one: even in the case of precise preferences, their findings would only have been able to influence individual behavior in a limited way, since single farms were very dependent on communal infrastructures and rooted in the statutory systems that held sway. This historically-generated "environment" also outlined the evolutionary path that could be followed—against naturally-imposed conditions if necessary. In the years after 1530, for example, a difficult-to-reach side valley in the Engadine in which hay had formerly been placed under shelter during the winter was opened to vehicular traffic. The investment of the commune in question became both possible and economically rational mainly because local farms already had carts and large barns in the village, which meant that it had already been centralized for the most part.[53]

In addition to the chronological constraints that forced each generation into a position of dependency on the preceding one, general economic conditions could also influence the ways in which space was organized. Among the important shifts were, as has been mentioned, the long-term increase in stable-feeding. At the end of the nineteenth century, one observer estimated that in the area near Belluno the amount of forage had more than doubled over the previous hundred years. Together with technological decisions, this created a dislocation in the existing system: thanks to road construction, there was less intermediate fodder storage on the high mountain meadows as larger quantities of hay were carried down in carts. In the valleys, greater need of space for hay storage near residential centers meant more and more haystacks either in the open or under simple shelters. Thus, the model of intermediary fodder storage (form 3) spread from higher to lower altitudes.[54]

It is significant that this process saw a substantial investment in transportation, while the built environment remained at a rudimentary level. In general, the conditions for investing in transportation and building construction began to change during the nineteenth century. Technical advances and increased demand helped create

Chapter Three

new traffic structures that altered the cost-benefit analysis of constructing multiple rural edifices. In this way, there was a shift in the previously-fundamental parameters of various developmental possibilities. The greater advantages now offered by centralized forms of transportation were indicated both by change in economic theory and, even more, by the very history of the period after 1900.[55]

In crop cultivation, erosion dangers made it more difficult to exploit steep lands than was the case with grass cultivation, since the vegetative cover did not provide permanent support, and since when the soil was being prepared it slid downhill. On the other hand, from a microclimatic perspective, slopes with good southern exposure were often well-suited to being planted with useful crops. Two kinds of labor investments were crucial for crop-growing under these conditions: the construction of terraces in order to reduce the steepness of the planted area, and the transport of soil in order to compensate for the progressive thinning out of the humus layer on uphill sections.

"One could say that two-thirds of these fields are artificial," wrote François Emmanuel Foderé in a scrupulous description of the department of the Alpes-Maritimes soon after 1800. He continued to explain that stacked-stone terracing was often inevitable in order to cultivate steep slopes and to protect them from the consequences of violent rainfalls. They used special tools to build these walls, including a metal lever with which to move large boulders. Inhabitants felt oppressed by the need to maintain and repair the walls constantly: "I've heard all the farmers claim that these walls are ruining them."[56] In our overall study area, though, fields on steppes that had been formed simply by frequent cultivation were more common than artificial terracing. Topographical conditions on these kinds of fields made a second kind of work regularly necessary: carrying soil uphill either on one's back or by other methods. Unlike terracing, this kind of work actually tended to increase the steepness of the given slope, in order to make sure that the upper portions of the fields had a humus layer that was deep enough. Evidence relating to such endeavors can be found in numerous regions. In the mountainous areas of the Val d'Aosta during the eighteenth century it was said that soil had to be re-transported uphill every three years.[57]

During the early modern period, in the Alpine space (and in surrounding areas) a great variety of tools were used for crop cultivation: hoes, spades, and plows with different shapes and pulling mechanisms. While in some areas this equipment was largely unchanged until after 1800 or 1850, in others there were considerable innovations. In the Dauphiné one finds seventeenth-century evidence of the diffusion of plows with a front support wheel, a previously unusual tool. In more and more areas of Styria and Carinthia plows with one-sided blades were commonly used, replacing the symmetrical spike plow (called the *Arl*). This change was not uniform, though. Along steep slopes the light spike plow, which also displaced less soil, beat out its competitor; it even remained dominant for a long while in central Carinthia, where the fields were mainly flat. The lay of the land was thus a factor that determined technology, but it was only one factor among others.[58]

These sorts of appeals to broad historical context could be reproduced for many places. According to a report from 1770, in a mountain community of the Vorarlberg all of the fields on the slopes had to be worked by hand with hoes "because they are much too steep," so that it was impossible to use plows. Even in the valleys cultivating with hoes was quite widespread, but not because of obstacles posed by the terrain: in the sixteenth century the fields had been plowed. Demographic growth, a diminution in property sizes, and agrarian intensification (in particular the introduction of maize) all led by the seventeenth century to a robust expansion in hand labor and to a decline in plowing. Taking into account the diffusion of processes of intensification, one can be sure that this development was not a special case.[59]

We have mentioned some of the elements that permit us to assess the role of technical factors in the development of Alpine agriculture. Until well into the nineteenth century transport technology was integrated into the existing organization of space and could influence this organization only to a certain extent. Not even plows with forward support wheels, or that could be used in more than one direction, provided across-the-board advantages. Although these had been known in part since the Middle Ages, their use spread only in some regions, while in other areas undergoing a progressive agrarian intensification, simple manual tools were employed. The marked dependence of technology on specific context points to its weak internal dynamic: agricultural development was determined until the end of our study period much more by the increase in labor inputs than by improvements in available equipment. International debates on agrarian technology spread everywhere, above all from the eighteenth century onward, and while they often painted a uniformly negative picture of traditional tools and celebrated newer ones, they had little impact on the ground. Increasing contacts of various kinds during this period broadened the potential for social innovation, but in practice many proposals for improvement were ignored. This was also true for areas outsides of the Alpine space. The technical revolution in agriculture did not really take place on a wide scale until the end of the nineteenth century, when industry began to compete seriously for labor and to produce equipment efficiently. There followed a quick transition to modern technology in some Alpine regions, such as in some parts of the French Hautes-Alpes, where population loss provided an extra incentive for this sort of innovation.[60]

On steep slopes, however, agricultural mechanization was particularly difficult. The new horse-drawn hay mower, for example, could only be used on relatively flat surfaces, even though in many places the quickness and ease of the hay harvest would have been especially important. As we have shown above, even during the early modern period steep slopes required far above average amounts of labor—what mechanization did was to amplify that pre-existing differential. And what is more, the increasing numbers of contacts and international discussions on agricultural improvement served to increase this differential in a subjective sense as well. Instead of contributing to regional agricultural development in the Alps, they served to sap peasants' enthusiasm for expanding their efforts. Perhaps even the perception of difficult mountain terrain became more acute. In any case, in the second half of the nineteenth century, one encounters drastically exaggerated popular expressions about steepness. At the time, people often used to say that the chickens of a given community must have iron spikes on their feet to keep them from falling down the hill, or that there, even the ants slid off the mountain. On the other hand, these kinds of expressions referred to places that were extremely steep, and thus to the fact that only a part of the Alpine space was composed of such communities.[61]

We can summarize the key points of this chapter as follows. Alpine agriculture was characterized by significant differentials of intensity within a small territorial radius. This was especially obvious in the difference between the extensive exploitation of Alpine pastures and the forms of intensive exploitation on the slopes near settlements. Extensive alpiculture provided large outputs per labor unit, but as a portion of total production, these amounts should not be overestimated. Most agricultural production came from the lower slopes, where livestock-raising, with its long-term stabling, was also particularly intensive.

Everything points toward the conclusion that during the course of our study period, the intensity differential was substantially consolidated. (1) At the beginning of the early modern period, large portions of the river valleys were exploited as pasture areas and waste lands from which naturally-occurring products could be gathered. Water control projects, first localized and then wide-reaching, caused many river bottoms to become the preferred areas for agrarian intensification. (2) Often harvest frequency increased more at lower altitudes than at higher ones. In some circumstances, peasants could collect as many as four harvests of fodder crops per year, and two harvests of comestible crops. Frequency increases on this scale were unproductive, or even impossible, in high-altitude areas with short growing periods. (3) Some plants and animals of recent introduction, or of growing importance, further contributed to these disparities. Corn production was limited, especially in the early phases, to lower altitudes. The increase in cattle with respect to sheep was tied to an increase in stable-feeding in the valley, while in the high Alpine zones pastures in favorable positions became more important.

Already during the early modern period, agricultural development had been conditioned by territorial factors in a selective way. But economic factors were also important. Even long before an increase in the agricultural output of Alpine land would have been unthinkable for contemporaries, huge labor requirements were already prohibitive in these territories. The amount of labor necessary for regular activities such as maintaining terraces and carrying soil uphill, or transporting people or goods to places that were inaccessible or territorially dispersed, can be well documented. The assessment of what constituted an acceptable amount of labor was made with respect to habitual forms of Alpine agriculture. It is certain that agronomic advances in the plains or in valleys during the late nineteenth century disinclined Alpine peasants to invest additional efforts. Prior to this time, technical factors had only had a limited influence on the development of Alpine agriculture.

In contrast to the pessimism pervading most studies of the agrarian potential of the Alpine environment, one finds here clear evidence of overall agricultural growth during the study period. Mountain agriculture was centered on the exploitation of lower-altitude lands near settlements, and demonstrated significant flexibility and adaptability in the face of population growth. Forms and techniques of cultivation show how demographic pressure was a crucial element of intensification processes. Irrigation, for example, led to a massive increase in hay production, but was very labor-intensive. The data indicate that the intensity of irrigation—due to demand and labor capacity—was very closely tied to demographic pressure. In the late eighteenth and especially in the nineteenth century, potatoes became an important crop whose output far outstripped that of traditional cereals as long as its massive labor requirements could be met. It was when the population grew that the disposition to produce the new plant increased, though potatoes had been available since the sixteenth century.

Chapter Three

Endnotes

1   On Malthus see Netting 1993 (citation from 278) and Viazzo 1989, 42–46.
2   For the national literatures, see *Österreich-Ungarn als Agrarstaat. Wirtschaftliches Wachstum und Agrarverhältnisse in Österreich im 19. Jahrhundert*, ed. Alfred Hoffmann (Munich, 1978); *Storia dell'agricoltura italiana in età contemporanea*, ed. Piero Bevilacqua, 3 vols. (Venice, 1989–1991); *Histoire de la France rurale*, ed. Georges Duby and Armand Wallon, 4 vols. (Paris, 1975–1976).
3   Frödin 1940–1941; an ethnological overview of traditional agriculture is provided by Niederer 1993, 147–224; the first agro-economic compendium was produced by Michel Cépède, see Emmanuel S. Abensour, *La vie rurale dans l'arc alpin. Étude internationale* (Rome, 1960).
4   Boserup 1981 and 1993; "labor productivity" is here used in its precise meaning of output per time unit of input.
5   Such is the case in the classic geographic study by Blanchard 1938–1956; a portion of the literature on agrarian history focuses instead on technical factors, taking as a model the unusual path of English development—see for example Paul Bairoch, "Agriculture and the Industrial Revolution 1700–1914," in *The Fontana Economic History of Europe* (Glasgow, 1973), 3: 452–506; the tendency to underestimate the importance of labor as a factor of production during the agricultural revolution has been criticized by scholars such as David Warren Sabean, *Property, Production and Family in Neckarhausen, 1700–1870* (Cambridge, 1990), 21.
6   Blanchard 1938–1956, 7: 344–345 and Jean Miège, "Inventaire des ressources agricoles de la Région Alpine," in *Economie Alpine. Actes officiels du Congrès de l'Economie Alpine* (Grenoble, 1954), 1: 119.
7   *Storia d'Italia*, ed. Giulio Einaudi, 6 vols. (Turin, 1972–1976) (for cultivated area during the eighteenth century, see 3: 544); Egidio Rossini and Carlo Vanzetti, *Storia dell'agricoltura italiana* (Bologna, 1987); Bevilacqua 1989–1991, as in n. 2.
8   A-Statistics 1861, 9. 62; Sandgruber 1978, 146, 148.
9   Ludwig Wallrath Medicus, *Bemerkungen über die Alpen-Wirthschaft auf einer Reise durch die Schweiz gesammlet* (Leipzig, 1795), iii–iv, 23–24.
10 Frödin 1940–41, (citation from 1: xxiii); methodologically he follows the work of Philippe Arbos, *La vie pastorale dans les Alpes françaises. Etude de Géographie humaine* (Paris, 1922).
11 For what follows see also Jon Mathieu, "Zur wirtschaftlichen Bedeutung des Alpwesens in der frühen Neuzeit," in Carlen and Imboden 1994, 89–104.
12 See table 2.3 above.
13 Mathieu 1992, 235; Frödin 1940–41, 2: 335–337, 515–518.
14 For the eighteenth and nineteenth centuries see Coppola 1989, 508.
15 Another point of view regarding the analytical utility of closed alpicultural models is provided by Viazzo 1989, 26–27.
16 Martonne 1926, 154–166; since he relied partly on older studies, his typology could also be said to refer to the period around the turn of the century.
17 Jan Caputa, "Potentiel de production agricole des sols d'altitude," in *Umbruch im Berggebiet*, ed. Ernst A. Brugger et al. (Berne, 1984), 245; Fritz Schnelle, "Beiträge zur Phänologie Europas II. 4 Mittelwertskarten: Gesamtvegetationszeit und 3 Vegetationsabschnitte," in *Berichte des Deutschen Wetterdienstes* 118 (1970); a synthesis on climate can be found in Josef Birkenhauer, *Die Alpen* (Paderborn, 1980), 174–203.

18   Christian Pfister, *Das Klima der Schweiz von 1525–1860 und seine Bedeutung in der Geschichte von Bevölkerung und Landwirtschaft*, 2 vols. (Berne and Stuttgart, 1985), esp. 1: 119–29, 143–51; the study cannot be generalized to fit the entire Alpine space without qualifications.
19   The notion of "frequency of cropping" was introduced into development theory by Ester Boserup (see n. 4).
20   Johann Jakob Staffler, *Tirol und Vorarlberg, statistisch, mit geschichtlichen Bemerkungen* (Innsbruck, 1839), 191–93.
21   Niederer 1993, 288–94; Clifford Thorpe Smith, *An Historical Geography of Western Europe before 1800* (London and New York, 1978), 525; for the frequency of hay-cutting, see Mario Romani, *Aspetti e problemi di storia economica lombarda nei secoli XVIII e XIX* (Milan, 1977), 404.
22   This increase can be best observed through regulations on the use of pastures and in land-improvement projects; see for example Nicolas Morard, "L'élevage dans les Alpes fribourgeoises: des ovins aux bovins (1350–1550)," in *L'élevage* 1984, 19–20; Ospelt 1972, 167.
23   Blanchard 1938–1956, 6: 421; *Historische Wasserwirtschaft im Alpenraum und an der Donau*, ed. Werner Konold (Stuttgart, 1994), 55–58, 149; *Actes du colloque international sur les bisses, Sion 15–18 septembre 1994*, special issue of *Annales valaisannes* 70 (1995). The intensification of livestock-raising was also an important factor (in the Valaisan example, especially during the fifteenth century).
24   Anton Gattlen, "Die Beschreibung des Landes Wallis in der Kosmographie Sebastian Münsters. Deutsche Ausgaben von 1544–1550," *Vallesia* 10 (1955): 140; F. G. Stebler, *Die Vispertaler Sonnenberge. Monographien aus den Schweizeralpen* (Berne, 1922), 81; Klaus Fischer, *Agrargeographie des westlichen Südtirol. Der Vinschgau und seine Nebentäler* (Vienna and Stuttgart, 1974), 73–74.
25   Fritz Schneiter, *Agrargeschichte der Brandwirtschaft* (Graz, 1970), esp. 56–91; Pickl 1982, 27–55; Karl Dinklage, *Geschichte der Kärntner Landwirtschaft* (Klagenfurt, 1966), esp. 77, 109, 114–16; Blanchard 1938–1956, 1: 305.
26   Thérèse Sclafert, "Usages agraires dans les régions provençales avant le XVIIIe siècle," *Revue de Géographie Alpine* 29 (1941): 471–92; Blanchard 1938–1956, for example 7: 307–10; Nicolas 1978, 2: 682–83; Paola Sereno, Les assolements biennal et triennal en Piémont aux XVIIIe–XIX siècles, in *Recherches de géographie rurale. Hommages au Prof. F. Dussart* (Liège, 1979), 357–68; on population issues, see table 2.3.
27   Mathieu 1992, 178; Johann Jacob Scheuchzer, *Beschreibung der Natur-Geschichten des Schweizerlands* (Zurich, 1706), pt. 2, 166; Hans Rudolf Schinz, *Beyträge zur nähern Kenntniss des Schweizerlandes* (Zurich, 1786), no. 4, 415.
28   Felix Monheim, *Agrargeographie der westlichen Alpen mit besonderer Berücksichtigung der Feldsysteme* (Gotha, 1954), 49–62.
29   Christian Pfister, "Bevölkerung, Wirtschaft und Ernährung in den Berg- und Talgebieten des Kantons Bern 1760–1860," *Itinera* 5/6 (1986), 367; Sandgruber 1978, 177. Examples for the early modern period in *Agricoltura e aziende agrarie nell'Italia centro-settentrionale (secoli XVI–XIX)*, ed. Gauro Coppola (Milan, 1983), 103; Nicolas 1978, 688–89.
30   The scholarship distinguishes between convertible farming (alternating between the production of food crops and forage) and permanent farming (separating fields and meadows spatially); the spread of the former in the northern Alpine zones is sometimes linked to the intensification of livestock-raising and consequent forage production, but the two forms of exploitation were linked by a host of factors and should not be understood generically as different stages of an intensification process.

Chapter Three

31 There exists no complete study of this topic, but see Peter Kaiser, "Das Wasser der Berge – Bedrohung und Nutzen für die Menschen. Notizen für eine Umweltgeschichte," in Bergier and Guzzi 1992, 54–108, esp. 86; one can find many examples in regional studies and in Konold 1994 (see n. 23); Blanchard 1938–1956.

32 Morard 1984 (see n. 22); Hans Conrad Peyer, *Könige, Stadt und Kapital. Aufsätze zur Wirtschafts- und Sozialgeschichte des Mittelalters* (Zurich, 1982), 156–94 (for ovines in the Alps, see 159 and also Hans Brugger, *Die schweizerische Landwirtschaft 1850 bis 1914* [Frauenfeld, 1978], 175, 178); the other interpretation is defended mainly by Fritz Glauser, cited along with other references in Mathieu 1992, 111. Differences of opinion can also be traced to the use of divergent criteria, and this is in turn tied to the unfortunate fact that different aspects of change are often not distinguished from each other: the structure of the array of livestock species, relations between livestock-raising and grain cultivation, and market-orientation vs. subsistence production.

33 *Geschichte der Land- und Forstwirtschaft* 1970–1980, 2: 587–89 (with tables in Slovenian); Ferruccio Vendramini, *La mezzadria bellunese nel secondo Cinquecento* (Belluno, 1977), 87 and id., *La rivolta dei contadini bellunesi nel 1800* (Feltre, 1972), 31; Coppola 1989, esp. 519–27; Placide Rambaud and Monique Vincienne, *Les transformations d'une société rurale. La Maurienne (1561–1962)* (Paris, 1964), 19, 23, 136–37; Viallet 1993, 235; for all of Savoy during the eighteenth century see Nicolas 1978, 2: 698–700.

34 For livestock numbers see *Atlas historique Provence, Comtat, Orange, Nice, Monaco* (Paris, 1969), esp. maps 221–222; Blanchard 1938–1956, 4: 343–44; *L'élevage* 1984, 271–276, 408–413; and in general *Histoire transhumance* 1986 (with bibliography).

35 The history of Provençal transhumance during the early modern period has been badly researched, but see the indications in Baratier 1978; Walter Panciera, *I lanifici dell'alto Vicentino nel XVIII secolo* (Vicenza, 1988), 73; decline in Italian zones does not exclude the possibility of the growth or even birth of transhumance in other regions.

36 Coppola 1989, 510–11, 526; Niederer 1993, 155–56.

37 For milk production see Felix Anderegg, *Schweizerische Alpwirtschaft. Illustriertes Lehrbuch* (Berne, 1899), pt. 2, 577; Staffler 1839 (see n. 20), 302; Bernard Bonnin, "L'élevage dans les hautes terres dauphinoises aux XVIIe et XVIIIe siècles," in *L'élevage* 1984, 275; Viallet 1993, 235.

38 Jean-Robert Pitte, *Terres de Castanide. Hommes et paysages du châtaignier de l'Antiquité à nos jours* (Paris, 1986), esp. 79–80, 95–97, 132–134, 199–201; Paola Sereno, "Sur les systèmes agraires originaux des Alpes Piémontaises. Observations de géographie historique," in *Les Alpes dans le temps et dans l'espace*, special issue (no. 125) of *Le Globe* (Geneva, 1985), 238–40.

39 Luigi Messedaglia, *Il mais e la vita rurale italiana. Saggio di storia agraria* (Piacenza, 1927), 48; *Geschichte der Land- und Forstwirtschaft* 1970–1980, 2: 576; Dinklage 1966 (see n. 25), esp. 139; Benedikt Bilgeri, "Der Getreidebau im Lande Vorarlberg," *Montfort* 3 (1948): 97.

40 Walter Brunner, "Frühe Nachrichten über Maisanbau in der Steiermark," *Blätter für Heimatkunde* 68 (1994): 5–15; Hans Telbis, *Zur Geographie des Getreidebaues in Nordtirol* (Innsbruck, 1948), 30; Messedaglia 1927 (see n. 39), 148–50, 276; Staffler 1839 (see n. 20), 205, 208; Hugo Penz, *Das Trentino. Entwicklung und räumliche Differenzierung der Bevölkerung und Wirtschaft Welschtirols* (Innsbruck, 1984), 206.

41 Viazzo 1989, esp. 183–192, 290–292; other authors have followed his tentatively expressed thesis.

42 Calculated by Sandgruber 1978, 155, 222 (for 1904–1913 and 1910) and Brugger 1978 (see n. 32), 16, 119 (for 1917 and 1910, likewise 1920); Switzerland is also an example of more dense potato

cultivation in areas surrounding the mountains; for clover in Carinthia see Dinklage 1966 (see n. 25), 175; for the beginnings of grass-growing for forage see also Blanchard 1938–1956, 5: 135; a late and limited introduction of these forage plants was typical.

43   The increase of crop yields for corn refers in particular to the seed/yield ratio; labor requirements for abundant corn yields is often underestimated by scholars, see Gauro Coppola, *Il mais nell'economia agricola lombarda* (Bologna, 1979), 56–75; various accounts suggest that for potato cultivation, labor requirements could be up to three times as great as for grain-growing.

44   Scaramellini 1978, 37–39 (citation), 58–63, 69–86, 165–80; Gauro Coppola, "Terra, proprietari e dinamica agricola nel Trentino del '700," in Mozzarelli and Olmi 1985, 719–27; Hermann Wopfner, *Bergbauernbuch. Von Arbeit und Leben des Tiroler Bergbauern in Vergangenheit und Gegenwart* (Innsbruck, 1960), pt. 3, 691; Bilgeri 1971–1987, for example 3: 192; Ingrid Zeller, *Weinbau in Vorarlberg* (Feldkirch, 1983), 11–29, 47–89; Staffler 1839 (see n. 20), esp. 221, 230, 234–36; Anne-Lise Head, "L'évolution de la typologie des zones agricoles en pays de montagne du XVIIe au XIXe siècle: définition et réalités du 'Hirtenland' dans le pays de Glaris," *Itinera* 10 (1989): 89–91.

45   Morard 1984 (see n. 22), 18; Peyer 1982 (see n. 32), 167, 177.

46   *Histoire Fribourg* 1981, see bibliography and chapters 7, 14, 21–23 (population levels for 1494–1811 on 280); CH-Statistics 1990, 6 (districts of Gruyère and Sarine 1798/1800–1900). Nineteenth-century agricultural science often focused on the relationship between sectors of production and intensification, see for example *Handbuch der gesamten Landwirtschaft*, ed. Theodor von der Goltz (Tübingen, 1890), 1: 348–53.

47   On the following question see esp. for southern Switzerland, Mathieu 1992, 117–62, and the ethnographic bibliography cited there, such as Anni Brockmann-Waldmeier, Richard Weiss, and Robert Kruker.

48   The typology refers only to the important factor of hay transport and stall-feeding; in reality, spatial organization was naturally also linked to other forms of exploitation (pasturing, plant cultivation, forest use).

49   Hans Rudolf Schinz, *Descrizione della Svizzera italiana nel Settecento* (Locarno, 1985) (a translation of *Beyträge zur nähern Kenntniss des Schweizerlandes* [Zurich, 1783–1787]), esp. 62, 78, 98, 248, 326, 378; for the *Heuzug* see for example the studies cited in n. 47.

50   References to the diffusion of these practices, esp. during the nineteenth and twentieth centuries, can be found in Niederer 1993, 63–64, 161–62, 174–75.

51   Paul Scheuermeier, *Bauernwerk in Italien, der italienischen und rätoromanischen Schweiz*, 2 vols. (Erlenbach-Zurich/Berne, 1943/1956), 2: 90–157, esp. 103, 114, 121, 135, 141, 147, 155–157; Scheuermeier points to the mountainous parts of Friuli, where characteristic baskets with wide weaves were frequently used, as another main zone of back-carrying transport.

52   Dinklage 1966 (see n. 25), esp. 147; on the Austrian passes see for example *Geschichte Salzburgs* 1983–1991, 1: 429, 610, 654, and 2: 2589; Rosenberg 1988, 20, 92, 97; Viallet 1993, 215–16.

53   Mathieu 1992, 128–32.

54   Antonio Maresio Bazolle, *Il possidente bellunese*, ed. Daniela Perco (Feltre, 1987), 2: 94–95, 116–17, 271; other general economic and social considerations could also play a role; many of the core areas of back-transport were for example in zones where cultivation of the soil by hand predominated (and thus draft animals were not employed) and where agriculture was on the whole based on women's labor (following male emigration).

55   Mathieu 1992, 130.

Chapter Three

56   François Emmanuel Foderé, *Voyage aux Alpes Maritimes ou histoire naturelle, agraire, civile et médicale du Comté de Nice et pays limitrophes* (Paris, 1821), 2: 2, 6, 25–26 (examples esp. of 1802–03).
57   Janin 1968, 156.
58   Vital Chomel, "Les instruments de labour traditionnels dans l'ancien Dauphiné," *Revue de Géographie Alpine* 52 (1964): 620–623; *Atlas steirisches Bauerntum* 1976, map 37; Dinklage 1966 (see n. 25) esp. 145–46.
59   Bilgeri 1949 (see n. 39), 19–25.
60   Blanchard 1938–1956, 5: 138–39; for a general look at the technical revolution in agriculture, see above, notes 2, 4, 5.
61   Robert Wildhaber, "Bildhafte Ausdrücke für Steilheit," *Schriften des Stockalper-Archivs in Brig* 12 (Brig, 1968): 103–109.

4 Cities

It has long been known that, both during the early modern period and in later times, there were few cities in the Alps. In 1588 the political writer Giovanni Botero, in his book *On the Causes of the Greatness and Magnificence of Cities*, stated that one reason why people gathered together to live in the same place was in order to seek protection against various dangers. Mountainous, harsh, swampy, or otherwise isolated places offered such refuge. Being Piedmontese, the phrase "luoghi montuosi ed aspri [mountainous and harsh places]" clearly referred for Botero to the Alpine ring that he had before his eyes. However, since the safety of these places was not tied to any particular benefits of territory, traffic, attractiveness or entertainment, "no very famous city was ever seen there." One can find similar observations, whether made in passing or explicated in depth, throughout our study period. Around 1730, for example, Aosta was described as a singular hybrid of city and village. Those who saw its chaotic peripheral districts with their dirty alleys, cows, and cattle stalls had to refer to it as a village, regardless of the Roman walls that surrounded the town. The rows of houses along the main streets in the center, however, gave the impression of a city; indeed, the quality of the construction was not bad "for a place in the mountains."[1]

The fame and obviousness of the modest level of Alpine urbanization has been an important reason for the lack of investigation in this area by modern research. Even geography, which has paid particular attention both to the Alps and to cities for quite

Chapter Four

a while, has only been looking at this problem for a few decades (when the urbanization of mountain regions could no longer be overlooked and thus became a problem). In the early 1970s a study of the several hundred towns within and just outside of the entire Alpine arc was completed under the direction of Giuseppe Dematteis. The work is interesting for issues of boundary-drawing and for a certain modernization of the image of the Alps, yet its historical analysis is scarce. Dematteis was satisfied with providing a summary retrospective overview and then concluding that the territorial distribution of the towns demonstrated that rough territory tended to resist urbanization, a general rule to which the Alps adhered. With a few exceptions, the results of other similar studies have been equally disappointing. Many geographers appeal in generic ways to elevation and the configuration of mountainous terrain in order to explain this 'stagnation' of urbanization. When urban development did occur (and this remains to be verified), it is attributed to transalpine transportation routes, though without detailed investigation.[2]

It is unsurprising that the historiography has not expended much effort to investigate a phenomenon that seemed obvious and independent of state activity. But historians have in fact been working on several aspects of this problem for quite a while, and since the sudden flowering of their interests in the postwar period, the research base has solidified substantially. The impact of systematic questions posed by geography has also enabled historical research on cities to develop in especially innovative directions over the past few decades. There have even been a few isolated efforts to establish connections between the fields of urban and Alpine studies, but these programmatic or essay-like contributions have failed to overcome obvious deficiencies.[3]

This chapter focuses on a specific part of this discussion. It offers a collection of urban demographic data and an analysis of observed growth phenomena, first for the early modern period and then for the nineteenth century. Many themes of modern urban history, such as the *fait urbain*—the specific urban effect or essence—will not be treated. Research on urbanization, which has to confront a similar question if only for organizational reasons, has still successfully made the claim that it is possible to examine urban reality from different points of view. This means that the traditional problem of how to define a city becomes somewhat less prominent. In what follows, I begin by conceptualizing the city in demographic terms, though I do not believe an urban area's sociopolitical position, and how others view that position, to be insignificant.[4] Given that an isolated view of the mountains could drastically reduce our results, the research must move beyond the Alps to include the surrounding areas.

Statistics in the early modern era

The most complete collection of demographic data for European cities during early modern times was made by Paul Bairoch and his group.[5] In this work the authors emphasized the difficulties involved in such a project. For many cities, demographic data for the period up to the nineteenth century are scarce and, for many reasons, imprecise. This inaccuracy increases with the interpolation of chronologically dispersed data into rounded-off years, which are useful for statistical representation. The quality of the statistics is linked to the study parameters that one establishes; incompleteness increases when one includes not only large cities, but also smaller towns, as did Bairoch, who set his limit at 5000 inhabitants. For the Alpine space, where cities were rare, just such an expansion of the parameters is crucial. Therefore, although Bairoch's data set can not for the time being give us a precise picture of the urban population, it can give us a sense of the order of magnitudes in the spatio-temporal realm that concerns us. This information is deployed here in order to give a picture of urbanization in the areas surrounding the Alps.

To be of use in examining the Alpine space, in a narrow sense, these data require clarification. In my view, the main problem with Bairoch's numbers is their tendency to fold outlying areas into urban concentrations. The population figures frequently refer to political territories that could also include secondary settlements. For example, Barcelonnette, a city in Haute Provence, is listed for the year 1750 as having 6000 inhabitants. In fact, Barcelonnette was really just a large community, whose nucleated center numbered only 2000 persons. The discrepancies in the case of Asiago, in the mountainous area of the Seven Communes, near Vicenza, were even greater. For 1850, the data set shows 12,000 inhabitants (thus including Asiago among the cities that prior to 1800 could possibly count 5000 inhabitants—though this was not proven by sources). A census carried out a bit later shows that the concentrated population of the "borgata" was only 500.[6] These observations led me to set at 5000 inhabitants the limit for inclusion as a city in the Alpine space, while also adding the criteria of a concentrated population of at least 3000 people. Thus we will examine the quantitative contours of urbanization in two ways: first, including the areas surrounding the Alps on the basis of Bairoch's figures; and second, limiting ourselves to numbers concerning the Alpine space that were taken from his data set, corrected, and even completed. This corrective work was based on an additional criterion and on specialized literature, but does not pretend to have produced numbers that are now exact.

Chapter Four

In the first case, the territorial demarcation is completely arbitrary. The two maps of the situation around 1500 and 1800 refer to the area between 4 and 18 degrees of longitude east and 44 and 49 degrees of latitude north, including the French coastal regions located farther south of this limit. This creates a space that reaches as far north as Karlsruhe (but just excludes Ratisbon), and as far south as the Appennines, with a total surface area of almost 580,000 square kilometers. To define the Alps, we use Dematteis' study cited above, which separates the mountains from surrounding areas according to morphological criteria. In order to minimize the subjectivity inherent in any attempt to create territorial boundaries, Dematteis identifies an intermediate zone between the Alps and neighboring areas. This area at the foot of the Alps extends from an idealized Alpine boundary line ten kilometers into the plain, and ten kilometers into the mountains. Only the cities located within this intermediate zone are considered Alpine cities. One could also object that the long, low-lying river valleys within the Alpine arc, such as the Adige valley, should be excluded, because such *golfes de plaine* ('flatland bays,' Emmanuel de Martonne) do not belong to the mountain zone. We will see below that this objection is linked to an interesting issue, but is only partially justified.

As one can see in map 4.1, the territorial distribution of the urban population in the study area was tremendously unequal at the beginning of the sixteenth century. In northern Italy there were famously many large cities, above all Venice and Milan, each one with about 100,000 inhabitants, while the regions north of the Alps were less urbanized. In addition to this clear north-south imbalance, there was also an east-west one. Within the Alps there appeared only one locality—the mining settlement of Schwaz near Innsbruck. Three centuries later, population in the same space grew substantially more dense, and cities began to spread themselves out more equally across regions (map 4.2). On the basis of Bairoch's data set, the increase took place in the following terms: 69 cities with a total of 1.2 million inhabitants in 1500, 97 with 1.7 million in 1600, 113 with 2.0 million in 1700, and 233 with 3.4 million in 1800. Since record-keeping is more troublesome for the earlier period, the increase was less pronounced, but the general tendencies, such as the push toward urbanization during the eighteenth century, can be realistically identified in the data.

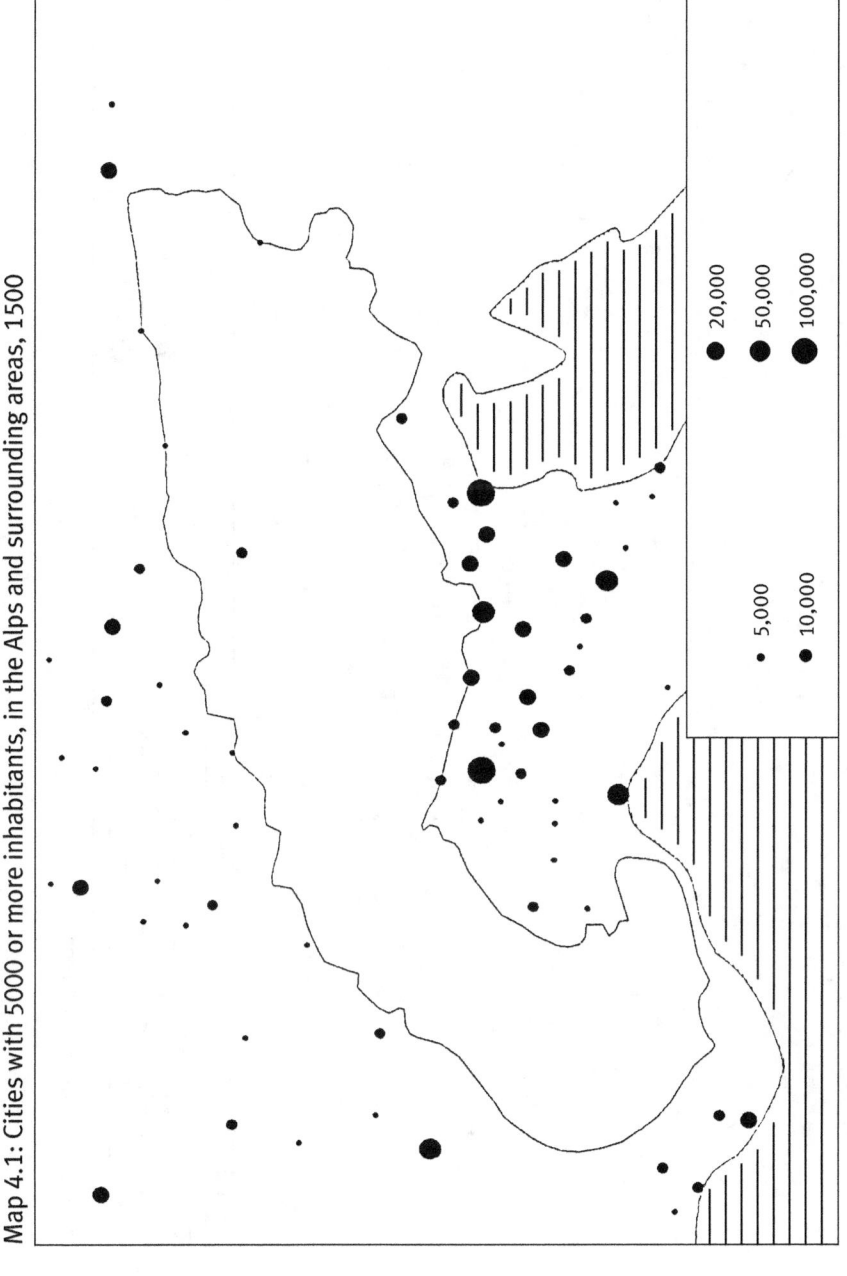

Map 4.1: Cities with 5000 or more inhabitants, in the Alps and surrounding areas, 1500

Based on data from Bairoch 1988, 4–69.

Chapter Four

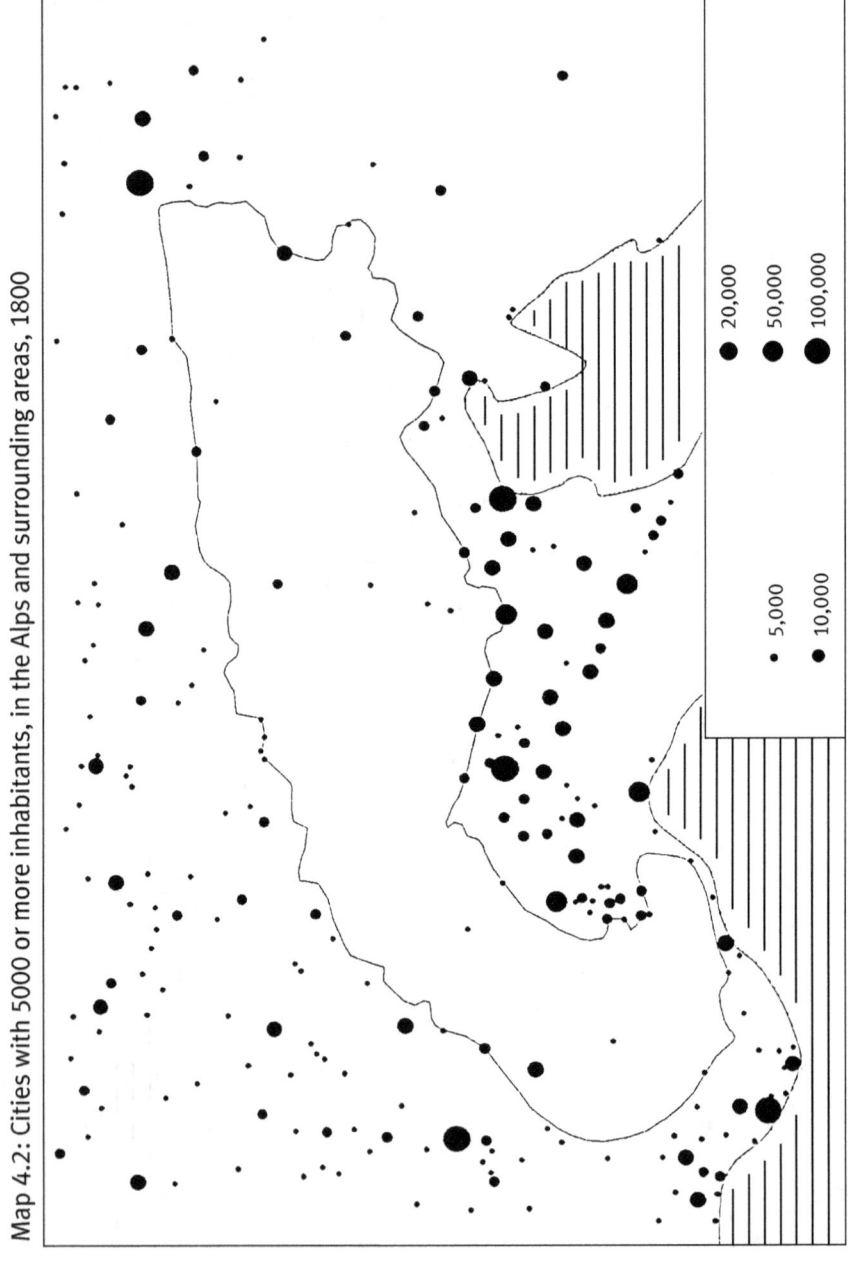

Map 4.2: Cities with 5000 or more inhabitants, in the Alps and surrounding areas, 1800

Based on data from Bairoch 1988, 4–69.

Cities

Table 4.1: Cities with 5000 or more inhabitants, in the Alps and at the foot of the Alps, 1500–1800

City	Inhabitants in thousands					Altitude (m)	Distance (km) to Alpine rim
	1500	1600	1700	1750	1800		
ALPS:							
Gap (F)	–	–	5	6	9	733	128
Grenoble (F)	2	12	20	23	20	214	15
Sisteron (F)	–	–	5	3	3	482	62
Aosta (I)	3	–	–	4	5	583	69
Belluno (I)	–	5	–	4	7	389	53
Bolzano/Bozen (I)	–	–	–	–	8	262	126
Rovereto (I)	–	–	–	5	8	205	44
Susa (I)	–	5	–	2	3	503	28
Trent (I)	4	7	6	9	11	194	68
Innsbruck (A)	4	6	7	10	12	574	106
Klagenfurt (A)	1	4	5	7	10	446	125
Schwaz (A)	17	9	8	6	4	538	78
FOOT OF THE ALPS:							
Annecy (F)	2	–	5	4	5	448	0
Chambéry (F)	2	2	10	9	10	272	0
Grasse (F)	–	–	9	9	9	333	0
Manosque (F)	–	–	5	4	5	387	7
Bassano (I)	3	7	–	8	10	122	0
Bergamo (I)	18	24	25	27	24	249	0
Biella (I)	–	7	–	6	6	420	0
Brescia (I)	49	36	35	30	32	149	0
Como (I)	10	12	13	14	15	201	0
Cuneo (I)	–	10	12	13	18	534	–8
Dronero (I)	–	–	–	–	6	622	4
Gorizia (I)	–	5	10	–	10	84	0
Ivrea (I)	–	4	–	6	7	267	0
Mondovì (I)	–	11	10	7	10	395	0
Pinerolo (I)	–	8	–	7	11	376	0
Saluzzo (I)	–	6	5	9	10	340	0
Verona (I)	50	45	35	45	51	59	–5
Lucerne (CH)	3	–	4	5	4	419	0
St. Gall (CH)	5	5	6	8	6	671	–5
Graz (A)	5	8	22	20	31	364	0
Salzburg (A)	7	9	13	15	16	424	0
Steyr (A)	6	9	6	7	8	307	–1
Kempten (D)	–	5	–	3	3	677	0
Maribor (SLO)	–	–	4	–	5	274	0

Cities listed have at least 5000 inhabitants (rounded to the nearest thousand, thus the list includes cities with 4500 inhabitants and more), at least 3000 of whom lived in the main residential concentration. City names and country indications are the current ones. Based on Bairoch 1988, 4–69 and on other sources listed below.

Chapter Four

Alpine boundary established according to morphological criteria; distance from the Alpine rim measured according to the most direct valley route; negative numbers refer to the distance to the rim from outside of the Alpine space, since 'foot of the Alps' is defined as a zone of 10 km on either side of the rim, based on Dematteis 1975, 84–99.

The following list indicates the completed data and corrections to Bairoch's data, displayed in the following form: city: Bairoch's data / comment and source, with complete citations given in the first instance. Corrections deal mainly with erroneous localizations and lowered population classifications based on the criteria of settlement concentration employed here. The numbers of inhabitants have been reported without alteration, with just a few exceptions: divergent data for a given year do not often strike one as more trustworthy, and interpolations in the rounding-off of the dates involve a consistent margin of discretion anyway.

Alps
Annaberg (D): coordinates for Annaberg im Lammertal (A) / refers to Annaberg-Buchholz (D). Aosta (I): – / inhabitants for 1500 and 1750 according to Lino Colliard, *Vecchia Aosta* (Aosta, 1986), 15, 211. Bad Ischl (A): 5000 inhabitants for 1800 / low level of settlement concentration, see *Bad Ischl. Ein Heimatbuch*, ed. Ischler Heimatverein (Linz, 1966), 226, 234. Barcelonnette (F): 6000 inhabitants for 1750 / low level of settlement concentration, see Jean-Joseph d'Expilly, *Dictionnaire géographique, historique et politique des Gaules et de la France* (Amsterdam, 1768), 5: 933. Gap (F): – / inhabitants for 1750 according to René Favier, *Les villes du Dauphiné aux XVIIe et XVIIIe siècles* (Grenoble, 1993), 437. Rovereto (I): – / inhabitants for 1750 according to Casimira Grandi, "La popolazione della città di Trento nel corso del Settecento: una capitale che si spegne," in Cesare Mozzarelli and Giuseppe Olmi, *Il Trentino nel Settecento fra Sacro Romano Impero e antichi stati italiani* (Bologna, 1985), 751. Schwyz (CH): 5000 inhabitants for 1750 / low level of settlement concentration, see Daniel Scheitlin, "Contribution à l'étude de la population urbaine suisse: 1200–1850. Constitution d'une banque de données et analyse des résultats," License thesis, University of Geneva, 1985, 2: 103. Susa (I): not indicated / inhabitants for 1600, 1750 and 1800 according to Karl Julius Beloch, *Bevölkerungsgeschichte Italiens* (Berlin, 1961), 3: 250; Raoul Blanchard, *Les Alpes Occidentales* (Grenoble, 1954), 6: 744; Giuseppe Prato, "Censimenti e popolazione in Piemonte nei secoli XVI, XVII e XVIII, *Rivista Italiana di Sociologia* 10 (1906): 345; René Le Mée, Population agglomérée, population éparse au début du XIXe siècle," *Annales de démographie historique* (1971): 500. Trent (I): 7000 inhabitants for 1800 / inhabitants for 1500–1800 according to Mariano Welber, "Due estimi e un principe. Trento prima e dopo il Cles," in *Bernardo Cles e il suo doppio*, ed. Mariano Welber (Trent, 1987), 168; Archivio Diocesano Tridentino, Visita ad Limina 1610, fol. 228 v; Grandi 1985, 744, 748, 761. Varallo (I): 5000 inhabitants for 1700 with coordinates for Varallo Piomba / refers to Varallo Valsesia with 3500 inhabitants for 1697, personal communication from Pier Paolo Viazzo, see id., "L'evoluzione della popolazione della Valsesia dagli inizi del '600 alla metà dell'800," *Novarien. Associazione di storia della chiesa novarese* 15 (1985): 121, 124.

Foot of the Alps
Altstätten (CH): 6000 inhabitants for 1800 / low level of settlement concentration, see Scheitlin 1985, 2: 6. Biella (I): 4000 inhabitants for 1800 / inhabitants for 1600 and 1800 according to Beloch 1961, 279; Le Mée 1971, 501 ("population totale"). Boves (I): 6000 inhabitants for 1800 / low level of settlement concentration, see Le Mée 1971, 501 ("enquête «1000»"). Busca (I): 5000 inhabitants for 1600 without source reference, 7000 inhabitants for 1800 / low level of settlement concentration, see Le Mée 1971, 501 ("enquête «1000»"). Dornbirn (A): 5000 inhabitants for 1800 / low level of settlement concentration, see *Österreichisches Städtebuch*, ed. Alfred Hoffmann (Vienna, 1973), 3: 115. Dronero (I): 3000 inhabitants for 1800 / inhabitants for 1800 according to Le Mée 1971, 501 ("population totale"). Herisau (CH): 5000 inhabitants for 1750, 6000 inhabitants for 1800 / low level of settlement concentration, see Scheitlin 1985, 1: 22 , 2: 51. Ivrea (I): – / inhabitants for 1600 and 1750 according to Beloch 1961, 279. Pinerolo (I): not indicated / inhabitants for 1600, 1750 and 1800 according to Beloch 1961, 279; Le Mée 1971, 500 ("population

totale"). Saluzzo (I): – / inhabitants for 1600–1750 according to Beloch 1961, 279; Prato 1906, 338. Bairoch also indicates a series of cities that probably—but without firm documentation—numbered 5000 or more inhabitants between 1500 and 1800. In the Alps: Asiago (I), Die (F), Digne (F), Lugano (CH), and Ormea (I). At the foot of the Alps: Caraglio (I), Giaveno (I), Gmunden (A), Neunkirchen (A), Verbania (I), and Vevey (CH). Studies have not produced evidence that any of these localities satisfied the criteria outlined above. Possible additions to our list include the Italian cities not indicated here: Ala, Arco, and Feltre (in the Alps) and Arzignano and Schio (at the foot of the Alps), which counted 5000 inhabitants during certain periods, but with a dubious level of settlement concentration.

During the entire period considered here, the towns at the foot of the Alps (those situated on the Alpine demarcation line in the two maps) experienced the slowest growth, going from 8 cities with a total of 150,000 inhabitants to 24 with 312,000. The decisive factor was not their position in the intermediate zone, but the fact that most of these cities were Italian. Growth in the heavily urbanized parts of northern Italy was much slower than in the rest of the area under analysis, especially in the seventeenth century, when northern Italy saw not only a slower growth rate, but an absolute decrease in numbers of cities and inhabitants. As if by irony, the Alpine cities were those that experienced the fastest growth during the early modern period: according to the same data set, they expanded from only one city with 17,000 inhabitants to 10 cities with 91,000. This disproportionate increase is the result of our statistical parameters, since we had fixed the limit of urban areas at 5000 inhabitants. Whenever a settlement passed that mark and entered into the urban category, it created a significant effect as long as the number of cities was small.[7] A more suitable picture is provided by the numbers related to urban density, which show that the absolute difference between the Alps and neighboring areas did not decrease, but increased notably. Around 1500 there was 0.1 city for each 10,000 square kilometers, and 0.1 urban inhabitants for each square kilometer in the Alps. In the surrounding areas the numbers were respectively 1.5 and 2.7. In 1800 these indicators increased for the Alps to 0.6 and 0.5, while for surrounding areas they reached 5.0 and 7.5. This difference widened at a particularly rapid rate during the eighteenth century.[8]

Table 4.1 contains the available population figures for the Alpine cities and for those at the foot of the Alps, organized by the countries in which they are currently located, from west to east. As has been noted, the numbers are not identical to those of Bairoch, which we have used up to now, but, where possible, they have been corrected, as is explained in detail in the appendix. Despite the additional criterion and the high degree of rounding off, we can not claim a great degree of confidence in these figures. It is thus difficult to decide whether, around 1500, there truly existed an Alpine city in the terms that we have been using. There may have been as few as none, or as many as five.⁹ The Tyrolean community of Schwaz demonstrated an anomalous structure and development pattern that do not really fit into the table (see below). Otherwise, the few cities that were clearly growing become apparent. In the first place is Grenoble, which after very rapid growth in the sixteenth and seventeenth centuries counted 20,000 inhabitants and easily overshadowed all of the other Alpine cities. It was followed by Trent, Innsbruck, and Klagenfurt, which in the eighteenth century reached the 10,000 mark. For every city in the Alps, there were two at the foot of the Alps, and these tended to be significantly larger than those within the Alps, thanks to the high population levels in Italian and, later, Austrian regions at the foot of the mountains. Surprisingly, there was little difference in the altitude of these cities. While the average elevation of a city at the foot of the mountains was 352 meters, that of an Alpine city was 464—this is not very high for a mountain chain with so many high valleys, and begs the question of what the conditions for urban expansion were.

Acceleration of growth

In order to identify specific causes of growth, it is prudent to begin with a general model of urbanization; that is, to assume that increased population density goes hand-in-hand with greater chances for urban concentration, and that activity in central locations changes according to a certain pattern of development. Variations from this pattern can then be identified as key factors for whatever case is being considered. It is not necessary, though, to attribute to individual cities unique urban functions, since this can easily result in artificial or even ideological categorizations. We will now look briefly at a few factors of acceleration that were either important for the development of Alpine cities, or that figure prominently in the specialized literature.[10]

"Urban reality in the Alps is linked to road traffic," states one monograph on the Val d'Aosta that looks at transalpine traffic after the fact, pronouncing explicitly a traditional opinion that remains widely held.[11] What can be safely claimed about this topic, on the basis of demographic data and research results concerning the amount of traffic and related occupational categories? If we compare cities located along the main communication routes—keeping in mind that transit along smaller valleys existed everywhere—then we find parallels for the Italian side of the Mont Cenis and the Montgenèvre (Susa around 1600) and for the Brenner route (Rovereto 1750–1800, Trent 1600–1800, Bolzano 1800, Innsbruck 1600–1800). A negative result is provided by the great passes that are today located in Swiss territory: around 1800 Chur counted 2500 inhabitants, while Bellinzona did not have even half as many. Only the figures for Lucerne (1750), a city at the foot of the Alps, could be situated in relation to the St. Gothard pass. In the second half of the eighteenth century, still, traffic increased massively on this and other routes, while the population of Lucerne decreased slightly. Susa and Innsbruck experienced parallel movements of increasing traffic and urban population, but their effective impact was weak. In the course of the eighteenth century, Susa's population grew to 3000, clearly smaller than it had been in 1600. It only qualified as an urban area again at the end of the nineteenth century, and this was *after* rail access from the plains to the new railway tunnel had detached the city from international traffic. Innsbruck experienced population growth during the sixteenth and eighteenth centuries that coincided with an increase in Brenner traffic, but it is not clear what happened in the crisis-ridden seventeenth century, when traffic stagnated.[12]

Records concerning occupational structure provide us with more certainties. Franz Mathis used tax rolls to show that the transportation trades were very weak in Innsbruck. In 1605, during the period when traffic was increasing, there were only nine independent wagoners working in the city, and two other individuals working in the transportation trade. A particularly detailed list from 1647 shows a total of nine independent operators in this field, amounting to 3% of active heads of household. One finds a similar percentage when employees are also taken into account. Generally, though, it is difficult to distinguish the transportation sector from related activities; one should also include the more numerous hostelry and other commercial businesses. On the other hand, many groups were not counted in the tax rolls at Innsbruck (numerous state officials, court servants, nobles, clerics), such that the small percentages mentioned above were based on a selection that only included about a third of the population. The transportation sector was significantly more important in the nearby town of Hall, where many goods arriving from the Brenner pass were transferred from the road to the river. This activity turned Hall into a veritable transportation hub, requiring consistently high volumes of labor. In 1647 no less than a fifth of the town's active population was involved in the transportation sector. But this center of traffic and of salt production did not become very large: in the seventeenth century it numbered 2500–3000 inhabitants and around 1800 barely 4000.[13]

In general, it can be said that none of the important Alpine cities was strikingly dependent on transalpine traffic. I would argue that this traffic did not even have that great of an impact in most of the cities with 5000 to 9000 inhabitants.[14] This assessment of the transit traffic contradicts the traditional view of many authors that grants a high profile to this sector, but it better fits recent research trends. The key points supporting this revision are: (1) the volume of traffic through these valleys was smaller during the early modern period than was often imagined from the perspective of the nineteenth and twentieth centuries; (2) a substantial portion of this volume was part of interregional, not transalpine, traffic (this should also be taken into account in the case of Hall); (3) for technical and organizational reasons, transport activity was mainly carried out by stages, with the result that economic potential was spread out along these routes rather than concentrated centrally.[15]

Location is also important when considering mining activities and their impact on urbanization. The role of the Alpine mining industry has been widely discussed in Austria, where many important extraction sites have existed since the Middle Ages. In a recent overview, Michael Mitterauer takes issue with older scholarship that gave mining activity a vital role in the creation of urban centers. He argues that the relationship between extractive activities and cities requires a more differentiated analysis, beginning on the economic level, since the impact of mining could differ completely depending on: the kinds of resources extracted (precious and non-ferrous metals, iron, salt); the stages of the labor process (extraction, preparation, transportation, commerce, final production); and the level of technology (which changed significantly over the early modern period). With respect to the extractive phase, mining precious metals required by far the greatest amount of labor. But even where mining activity generated sizeable population increases, this translated into an urban concentration only in certain circumstances: "In the areas around extractive sites, mining activity penetrated into the landscape of rural settlement. It led not so much to the formation of single centers, but to the creation of territorially expansive districts." Still, Mitterauer does not deny the possibility that within these districts industrial or market centers could sometimes form or increase in size.[16]

A perfect example of a mining center is Schwaz, which makes a spectacular opening in our analysis of large Alpine cities (table 4.1). In effect, this settlement experienced an extraordinary boom as it passed into the early modern period, reaching its peak at the beginning of the sixteenth century. The mining industry of Schwaz was at the time the leading center of European copper and silver production. In 1523 about 17 tons of silver and 1400 tons of copper were extracted from the main mine. But Schwaz was actually comprised of three distinct mining districts in which at least 9000 workers were active during the 1520s, 30s, and 40s. In Schwaz itself a series of churches and large secular buildings, including a palace of the Fuggers, merchant-bankers from Augsburg, was constructed at the turn of the century. What remains very uncertain about the history of Schwaz is how many people lived there. Recent work has shown that the primary sources do not confirm many of the overestimated figures that have been cited in other studies. The relatively densely populated central settlement might have had only 300 to 350 houses with perhaps 3000 inhabitants in the middle of the sixteenth century. Although later both mining activities and surrounding districts suffered a sharp recession, the center appears to have continued to develop into the eighteenth century, until it also experienced a reversal. Thus, when one takes the densely populated, properly urban zone into account, developments over the long term seem almost opposite to those suggested by our data.[17]

As historical research has indicated, sociopolitical factors were often very important in the creation of large cities. In Klagenfurt, a particularly instructive example, the push came from the regional nobility, which developed a group identity during the fifteenth century and began to meet more and more frequently in regional assemblies in Carinthia. Continuous warfare along the southern borders, several peasant revolts, and also a desire to create some autonomy for themselves with respect to the Hapsburg prince (following the example of other regions), led this aristocratic union to seek a fixed location. Their choice was not St. Veit, a city that already functioned in some ways as a center, but nearby Klagenfurt, which was less important. Klagenfurt numbered about 800 inhabitants and had recently been devastated by a fire. In response to their request, the emperor handed the city over to the noble estates in 1518. They created permanent institutions and began to lay the groundwork for a regional administration. At the same time, the lords of the city began an expansive urban reconstruction project. By about 1600 the city's profile had totally changed, thanks to a canal linking it to Lake Wörther, the Italian-style fortifications, the *Landhaus*, and other representative buildings. In the meantime its population quintupled, in large part due to the growth of the local craft industry. In the eighteenth century it seems that manufacturing expanded again. Its presence was now larger, more differentiated, and partly oriented toward new market sectors. Benefiting from protectionist policies, some large companies also began to produce for the international market. But Klagenfurt remained a city in which nobles, officeholders, and clerics counted for a large percentage of the population—half, according to a document from the early nineteenth century.[18]

Nobles also played an important role in Grenoble, by far the largest city in the Alps since the seventeenth century, but they had long been allied with the king of France and found themselves under his increasingly strong and direct influence. The appellate court of Dauphiné, created in the fourteenth century and soon thereafter transferred to Grenoble, was recognized in 1453 as a royal *parlement*. It enjoyed an extraordinary number of jurisdictional privileges, including the right to convoke the noble estates of this large province, which stretched from the Rhône valley into the mountains. Until their dissolution, Grenoble also developed as the center of the noble estates assembly. During the sixteenth century, the provincial governors oversaw new forms of military organization. The most powerful of them, François de Bonne, duke of Lesdiguières, had an urban residence built for himself in around 1600 and began to expand the city center according to the standards of the times. When royal intendants finally arrived in the province later in the seventeenth century, it was obvious that they would reside in Grenoble. As René Favier explains in his discussion of the capital city and the other centers of Dauphiné, this successive concentration of authorities made its mark.

The development of administrative functions concentrated in the city an exceptional number of legal personnel and royal officers, and, beginning in the early seventeenth century, the majority of the provincial nobility. Compared to the modest centers inhabited by rentiers, the artisanal burgs, and the minority of towns where elite merchants competed for preeminence with the lawyers and royal officers, Grenoble was the only town in Dauphiné that really attached itself to an aristocratic model.

The fact that these pressures came from the aristocratic elite does not mean that this group and its hangers-on accounted for the majority of the population. In the period of Catholic reforms the clerical rank was expanding, and in the eighteenth century the city was also the seat of an important military garrison. But above all, industry and commerce had developed and diversified energetically, as soon as urbanization had started to take off. Glove manufacturing, geared in the eighteenth century toward markets that reached overseas, was particularly successful and provided employment for many workers in Grenoble and its surroundings. However, population growth in the city slowed somewhat from 1700 onward.[19]

The important influence of sociopolitical forces on urban growth in the Alps is also demonstrated by other cases, especially that of Innsbruck, occasional residence of the Hapsburgs. Further, a kind of reverse proof of the impact of these forces is provided by the fact that no city in the mountainous areas of the Swiss Confederation can be classified as urban according to our parameters. If the transit traffic had supplied the sorts of impulses for urban development that has frequently been attributed to it, then one would have consistently found numbers of urban concentrations in this area. In fact, the influence of this traffic on city growth remained modest, whereas the development of state and society played an important role therein. In the case of the Swiss Alpine regions, both of the latter retained a localized character, blocking the possibility of accelerated urbanization.

Even though the factors mentioned here had varied impacts, each of them contributed to urban growth according to specific regional conditions. This was also true for many other parts of Europe during the early modern period.[20] In what follows, we will examine the relationship between the Alpine environment and urbanization—where one may expect to find the peculiarities of the topic that concerns us. This leads us first to a consideration of the possibilities and limitations of Alpine agriculture.

The slowing of urban growth

Two economic factors are usually considered to be fundamental for urban supply systems: the production of an agrarian surplus, and the availability of means of transporting it. While the second factor does not lend itself to theoretical debate, the question of the conditions under which a surplus could be created is a controversial one. Paul Bairoch approaches this issue from the perspective of the level of agrarian technology. He argues that in traditional situations, both the productivity of an agricultural worker and the potential for producing a surplus were limited, creating a specific ceiling for urban growth that could only be surpassed following a revolution in agrarian technology. Ester Boserup, on the other hand, shows that the crucial element was not the size of the surplus produced by a single worker, but the total available surplus in a given region. Depending on population density and agricultural intensity, the total available regional surplus could vary significantly, even in pre-industrial circumstances. In this way, urbanization depended on demographic growth rather than on technological factors.[21] Contrary to what one might think, this second theory is particularly useful for our investigation: instead of beginning with a problematic model of a fixed ceiling for urbanization, it focuses on the intensity differentials at comparable technological levels. This is well-adapted to the differential that one finds between the Alps, where most settlements were isolated and thinly-populated, and surrounding areas, which were in part densely-inhabited and urbanized.[22]

How did the provisioning of early modern urban centers take place? Existing studies have for the most part looked at the lowlands and make clear the difficulty involved in the detailed description of even a single case. One must take into account the great variety of ways by which supplies reached the city: urban households sustained themselves by cultivating gardens and other plots both inside of the city and in the immediate surroundings. Landowners or those with jurisdictional powers of various sorts collected revenues in kind from nearby lands. Other kinds of foodstuffs arrived in the city through commercial circuits, whether at the weekly markets where peasants sold their own products, or through merchants and salespeople. From the late Middle Ages on, many aspects of urban provisioning fell subject to statutory regulation, producing very useful records—but ones that must nevertheless be read carefully if one is to avoid unrealistically rigid notions of economic processes. Exchange had to be particularly flexible during crisis periods, when it was necessary to secure food from sources far beyond the region. Even in normal periods, the areas that provisioned the town differed according to the product in question. Researchers have focused by and large on the supply of grain. The specialized literature has established a standard zone of provisioning for an early modern city of 20,000 at a radius of 15–20 kilometers from the center, positing that a city could survive on the food produced within this area.[23]

However the parameters are fixed, they must be altered for the Alps. The most important factor limiting the agricultural potential of mountain areas was the vegetative period, which grew shorter as elevation increased. This reduction resulted in the production of smaller amounts of forage, and in a lower likelihood of fields being sown a second time after the main harvest. That being said, it is not necessarily the case that high elevation and agrarian intensification were mutually exclusive, though output increase for a given surface area either required more labor or was simply lower—an important condition as far as the provisioning of cities is concerned. Hans Bobek, one of the founders of functionalist urban geography and author of a classic study of the development of Innsbruck, drew attention to the restrictive conditions posed by the territory surrounding the city.[24] Within the so-called one-day zone, that is, the area within a 15-kilometer radius that could be reached by foot in a round-trip of one day, two thirds of the land was over 1000 meters in altitude, whereas the lower-lying settled area was restricted to the remaining third. Further, the distance of 15 kilometers (normally a four-hour walk) was not a reliable standard, since the terrain varied tremendously at different elevations. When the variability of surface area is taken into account, the size of the 'one-day zone' shrank considerably (see map 4.3).

As has been pointed out, the average elevation of early modern Alpine cities (those with at least 5000 inhabitants) was about 460 meters, only 110 meters higher than medium-elevation cities at the foot of the mountains. What clearly distinguished the two groups were not the altitudes of the centers, but those of the cities' immediate surroundings. If one takes the 15-kilometer radial area and a 1000-meter elevation as parameters, one finds that 31% of the outskirts of Innsbruck was lower than 1000 meters. In two Alpine cities on our list, the percentage was notably smaller (17% for Aosta and 16% for Susa). But besides these and other places with very restrictive conditions, one finds a series of cities with percentages of 50% to 75%. Yet even here differences with respect to cities at the foot of the mountains are important. The cities situated along the Adige River provide a good example. For Verona, at the foot of the Alps, 100% of the outskirts are located below our altitude parameter. Only a short distance upriver, moving into the mountains near Rovereto and Trent, does the percentage decrease by a third. At Bolzano, the figure is barely 50%, though the city is located at an elevation of only 260 meters. The potential importance of an urban center's general location is demonstrated by the two cities of St. Gall and Kempten, on the northern edge of the Alps. They are higher in elevation than most Alpine cities, but almost 95% of their outskirts are below 1000 meters. There is one exception in our group of Alpine cities: at Klagenfurt almost all of the immediate surroundings are lower than our parameter. The favorable terrain must be one of the reasons for the urban density that is attestable from an early date in the Klagenfurt basin.[25]

Another factor that can be observed fairly systematically is the transportation situation. River traffic on boats (not transport by rafts or by floating, that can be carried out only downstream) is an indicator that lends itself to analysis particularly well. Transportation by boat on rivers and canals was slow but inexpensive, and it was an important means of provisioning large cities during the early modern period, especially during the sixteenth and seventeenth centuries, because its economic advantages became more apparent as the distances involved increased. Although river navigation was in part a man-made resource, as demonstrated by the significant obstacles to such traffic, the Alpine arc was at a disadvantage in many ways compared to surrounding areas. In the mountains boats were smaller, navigation was more difficult, and there were many waterways that simply could not accommodate traffic, at any amount of effort and expense. Within the Alps one can posit some sort of relationship between the size of cities and river traffic. Of the twelve cities on our list, eight—a substantial number—were linked, directly or quite nearby, to river traffic. Among these were the four largest cities, which numbered 10,000 to 20,000 inhabitants at the end of the eighteenth century, while the largest of the cities without river traffic were much

Chapter Four

Map 4.3: The area surrounding Innsbruck, 1928

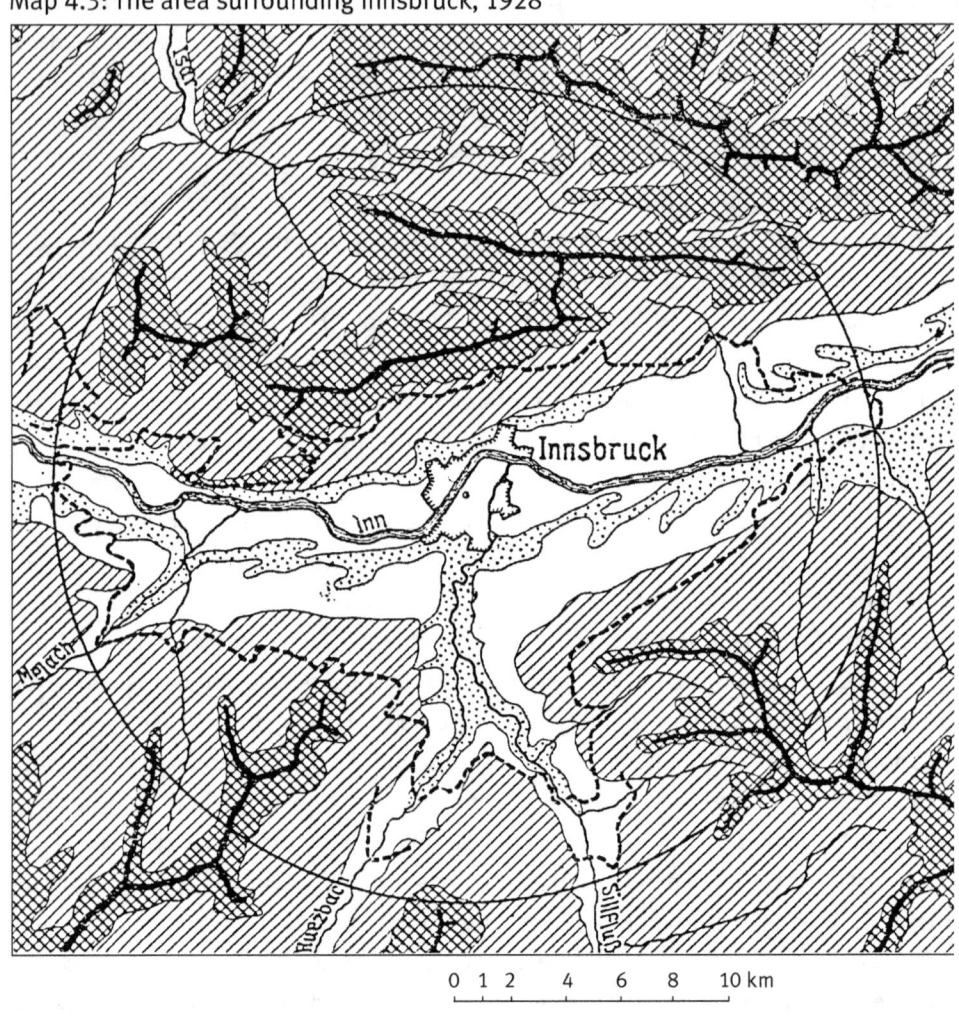

	area conducive to plant cultivation, generally under 1000 meters
	mainly forest and pasture areas, above 1000 meters
	under 1000 meters, plant cultivation impractical due to steepness; mostly forest
	mountain peaks, rocky and non-productive terrain

The circle has a radius of 15 km

4-hour isochronometric (reachable by foot or by wagon) ---

Source: Hans Bobek, *Innsbruck. Eine Gebirgsstadt, ihr Lebensraum und ihre Erscheinung* (Stuttgart, 1928), 87.

smaller: Belluno then had at most 7000 inhabitants, and Gap did not have many more than that.[26]

In order to illustrate other aspects of the urban economy, we will pause briefly to consider the last two places mentioned. Belluno, in the Venetian *Terraferma*, found itself in a relatively favorable position for an Alpine city, and the same could be said for its hinterland, where many of its citizens owned a considerable amount of property. Despite this fact, there were continual conflicts over provisioning. The rural communities, for example, submitted a complaint to the Venetian state chancery in 1580 about being coerced in their market activities. They argued that they were being placed in an impossible position by being forced to sell firewood in the city during the winter months, since there was often not enough for their own communities. They were also upset about the city council's insistence that every other week 300 persons, according to a schedule that rotated among the communities, had to supply chickens, cheese, butter, and other goods at a dairy market in the city. Given their many other activities and work, the complaint read, it was inevitable that they would be unable to meet these requirements and subsequently fined, "since many of the villages were 10 or 12 miles, or even farther away from the city." In fact, the burden of supplying the market fell upon all of the villages within a radius of about 15 kilometers, while a community a bit farther out in the same district was exempt. We do not know exactly how many peasants lived in the city of almost 5000 inhabitants, but there were surely more than a few. In the late eighteenth century they still accounted for almost 10% of the active population.[27]

It has been documented for the entire Alpine arc that peasants were a constant presence in early modern cities, but their numbers and profiles were subject to strong fluctuations. This can be partly explained by the sizes of the cities, and should also be situated with respect to factors that are more difficult to define, such as cultural *habitus* or the prior history of the region in question. In southern French cities agricultural workers formed a very high percentage of city inhabitants. In Gap it is estimated that the 'rural' population of the city during the Old Regime accounted for a good 60% of the total. Within the city walls one found numerous day laborers, working mainly in agriculture, whereas a great many farm owners, leaseholders, and servants lived in the immediate surroundings. Thus, the urban commune was practically able to be self-sufficient. For 1730, for example, annual cereal production was estimated at 185 kilograms per inhabitant. According to the consumption standards of the time, an

Chapter Four

output of this amount indicates that the city was 95% to 126% self-sufficient—surprising figures given that of all of the cities on our list, Gap's elevation was the highest, and its surroundings were not particularly favorable. One must take into account, however, the city's extensive area (110 square kilometers) and the fact that its high percentage of agricultural workers meant that there was only a small non-agricultural sector that had to be provisioned.[28]

The situation in Grenoble was completely different. In this large city agriculture was of secondary importance, and provisioning was strictly regulated, drawing from a vast supply area. Here during the second half of the eighteenth century the municipal hospital drew some of its grain from the valleys of the Drac and the Isère, but most from the densely-inhabited plain of Bièvre situated outside of the mountains. Almost all of the grain at Grenoble's market came from distances of 30 to 50 kilometers. The fact that the city's immediate surroundings contributed only modest amounts was also related to the fruit orchards and the hemp production that were spreading out along the Isère in response to urban demand. These forms of agriculture, carried out with much fertilizer and without fallowing, were among the most intensive in the entire province at the end of the eighteenth century, and were praised as a model of agricultural progress: "Would that an agriculturist or a chemist come from the banks of the Seine to take a glance at the harvests of the valley of the Isère; would that they compare the fruits of their soil to our hemp, which are sometimes ten to fifteen feet tall; would that they carefully examine to see whether there is a single square foot of land that is not under cultivation or in productive use!"[29]

One can observe phenomena linked to intensification, both on large and small scales, in the vicinity of many Alpine cities. Relationships between the Alpine environment, population and urbanization were thus not one-sided. Once urban centers emerged and developed, the increased demand that they created had an impact on the population and land use of the city's surroundings. These effects in turn generated new needs to which the city responded. As altitude increased however, these densification processes became less and less likely. Since we only have sporadic evidence for the early modern period, we are unable to compare cities in this respect to the entire settled area of the Alpine arc. To help us around this problem, we will rely on a modern-day collection of the altitudes at the settled center of all of the communal territories in the Alps (the measurement point is not that of the cultivated areas since these are often at higher elevations). Even though, for chronological and methodological reasons, the source of these data privileges lower-altitude areas,[30] the comparison points nonetheless to significant differences between the cities analyzed and the total amount of communal territory. In around 1800, two thirds of the cities and their populations were situated at elevations below 500 meters, even though this altitude level did not account for even a quarter of the total surface area of Alpine communes (table 4.2).

Table 4.2: Cities and urban population in the Alps according to altitude, 1600–1800

Altitude (m)	1600		1700		1800		Alpine area (1990)	
	C	I	C	I	C	I	km²	%
0–499	3	24,000	4	36,000	6	64,000	41,100	23
500–	3	20,000	3	20,000	3	26,000	92,300	51
1000–	–	–	–	–	–	–	40,400	22
1500–	–	–	–	–	–	–	7,700	4
	6	44,000	7	56,000	9	90,000	181,500	100

C = the number of cities with 5000 or more inhabitants, I = the number of inhabitants in these cities.
'Alpine area' incorporates communal territories classified according to altitudes of communal centers in 1990, see Bätzing 1993, 75. Sources and criteria for the cities as in table 4.1.

The table also shows us that the number of cities and the total urban population located at the lowest elevation level doubled between 1600 and 1800, while those at medium levels experienced only a slight increase. This can be seen as still more evidence linking agricultural intensification to urbanization during the early modern period, especially if one takes into account the fact that behind the relatively unremarkable altitude differential of the urban centers one can find a much more prominent differential in the outskirts of urban areas.[31] We must stress again, however, that our sample group is too small to make statistically verifiable claims, and that it is based on incomplete and imprecise data. Some other observations are at least as interesting. When in 1746 the French military administration calculated the expenses submitted for the provisioning of a battalion posted in various garrisons in the Dauphiné, Briançon, located at an elevation of 1300 meters, was the most expensive. Its costs were 40% higher than those estimated for the cities in the Rhône valley. It appears that in mountain cities it was particularly costly to meet large additional demands on the food supply, because usually the price levels in the Briançonnais were barely different from those in the lowlands. These problems were among the reasons why the French strategists grew so disillusioned about the fortress at Montdauphin. The plans for this fort, whose construction along the mountainous border with Savoy-Piedmont began

Chapter Four

in 1700, included not only a strong garrison, but also the expectation that a large civilian population would follow and create a true urban center. But history took another direction: what was hoped to have become a pre-planned Alpine city of the eighteenth century remained instead a small presidio.[32]

The nineteenth century

In the nineteenth century, statistical production entered a new phase. Together with the formation of national state administrations, especially after mid-century, it underwent an enormous development while also achieving a certain level of international standardization. Thus, from 1870 onward, we have simultaneous, precise, and detailed census data for our entire study area. As our point of departure we will take the data from the years around 1870 and 1900, which reflect a particularly vigorous phase of urban growth.

Since quantitative record-keeping became substantially more precise than in the previous period, the problem of criteria now becomes more urgent for us. We will continue to measure the urban population in terms of the territorial definition of the time. Among the alternative methodological options, one is already eliminated because of the nature of the documents: the concentrated population, which is that which really interests us, is not differentiated from the total population in all of the censuses. A second method that is used in quite a few studies would be to employ the communal boundaries as they exist today. However, creating this kind of a retrospective statistical set is disadvantageous due to the many cases of communal incorporation during the twentieth century, which if projected into the past would produce completely unrealistic results. In comparison, distortions stemming from use of the variable definitions created during the study period itself are modest. It is true that in some cities, political expansion significantly preceded population growth (or vice versa), but these are exceptional cases whose impact on the results is limited.[33]

What distinguishes the data collected in the late nineteenth century from those of the period up to 1800 is their systematic and reliable character. I believe that the two data sets will lend themselves to better comparison if we continue to use our parameter of 5000 inhabitants as the threshold for classification as a city, while setting more restrictive definitions for additional criteria. This applies to the level of population concentration (which must be defined in some cases by using lists of localities) and to our guidelines for rounding off numbers (see table 4.3). Despite the more restrictive definitions, the statistics show considerably accelerated growth in the number of cities and the size of the urban population. The numbers more than double, first between 1800 and 1870, and then again between 1870 and 1900, and there is a fivefold increase over the course of the whole century.[34]

Table 4.3: Cities and urban population in the Alps according to size, 1800–1900

Inhabitants	1800		1870		1900	
	C	I	C	I	C	I
5000–	2	12,000	7	41,500	18	109,000
7500–	3	25,000	7	57,800	10	83,600
10,000–	3	33,000	1	13,100	8	92,800
15,000–	0	0	4	64,700	2	34,300
20,000–	1	20,000	0	0	3	76,100
30,000–	–	–	1	42,700	0	0
50,000–	–	–	–	–	1	68,600
	9	90,000	20	219,800	42	464,400

C = the number of cities with 5000 or more inhabitants, I = the number of inhabitants in these cities.
Sources and criteria for 1800 as in table 4.1. For 1870 and 1900 parameters are tighter (rounding-off to 100, thus precisely 4950 or more inhabitants, at least 3500 living in the main settlement); the detailed list with source references is in the appendix, table A.6.

This growth was especially prominent in the small cities that passed the urban threshold in the late nineteenth century. Among these one found a new phenomenon for the Alps: suburbs of places that already qualified as cities, such as Hötting and Wilten just outside of Innsbruck. In 1900, Innsbruck had 26,900 inhabitants, and if one included the two suburbs the number would have reached 45,100. But even if one were to count like this, Grenoble would remain the largest city by far, over the course of this entire period, first with 20,000 inhabitants, then with 42,700, and finally with 68,600. Generally, city growth was closely linked to starting size. This was already true for the period between 1800 and 1870, and it can be even better documented with the larger sample of 1870. The annual growth rate of cities with 5000 to 10,000 inhabitants averaged 5.6 ‰ until 1900. For cities with 10,000 to 20,000 inhabitants it was 12.2 ‰, and for Grenoble it was 16.5 ‰. The fact that larger cities had better growth possibilities than smaller ones does not mean, however, that they grew faster than all other places. The highest growth rates are associated with those small communities that crossed the urban threshold only between 1870 and 1900. Two examples are Knittelfeld in Styria and Davos in the Grisons, each of which had 8100 inhabitants in 1900, and whose annual growth rates for the period were 46.2 ‰ and 47.7 ‰, respectively.

Chapter Four

For documentary reasons, it is easier to relate the history of urbanization in the nineteenth century to general demographic history than it is for earlier times. During the early modern period, this relationship could only be sketched in broad strokes. The vigorous urban growth of the sixteenth century, and the decline and subsequent renewed growth in the seventeenth and eighteenth centuries correspond to the overall trends in Alpine demography. One can now view the evidence of the very rapid nineteenth-century urban expansion against the background of population increases that created in many places—though not everywhere—a huge break with Old Regime patterns.[35]

These regional differences were very important for the changes in the territorial distribution of cities. If we start from the basis of the countries in which they are located today, then in around 1800 the French and above all Italian Alpine spaces seriously outweighed other areas with respect to cities and urban populations. Between 1800 and 1870 the Swiss regions experienced an especially rapid demographic expansion and an increase in urban density. The Austrian regions remained at first a bit behind, but then caught up in both respects between 1870 and 1900. Still the fact that in the entire Alpine space the rate of urbanization was higher than the general rate of growth means that the relationship between the two is of limited explanatory value. For the French regions one finds an especially large gap: there, the cities grew just as fast as elsewhere in the Alps, but the regional population remained unchanged, or even decreased. This shows how important national influences were, because this kind of gap could be found in many parts of France.[36]

Significant changes in transport systems were necessary requirements for the opening up of the urban economy that was characteristic of the nineteenth century. New road construction, which had already accelerated during the Old Regime and then continued by force during the Napoleonic period, combined with technical and organizational innovations to create notable changes by the middle of the century. The first rail lines reaching Alpine localities were opened during the 1850s. By about 1870 the large centers were linked to developing networks, and by about 1900 almost all of the cities that have been discussed here had rail links. In general, one must assume that it was now easier for towns to extend their supply systems beyond the hinterland. The internationalization of cereal markets also helped here: by the 1870s the imbalance of supply over demand led to a readjustment of food prices. But in individual cases, the transport revolution did not guarantee accelerated growth. Susa, which had had a rail connection with Turin since 1854, displayed a very modest growth rate of only 4.3 ‰ during the last three decades of the century. Alongside these kinds of examples, though, there was also a series of cities for which the railway became an important growth factor.[37]

Table 4.4: Cities and urban population in the Alps according to altitude, 1800–1900

Altitude (m)	1800		1870		1900		Alpine area (1990)	
	C	I	C	I	C	I	km²	%
0–499	6	64,000	12	153,500	20	268,300	41,100	23
500–	3	26,000	8	66,300	20	180,600	92,300	51
1000–	–	–	–	–	1	7,400	40,400	22
1500–	–	–	–	–	1	8,100	7,700	4
	9	90,000	20	219,800	42	464,400	181,500	100

C = the number of cities with 5000 or more inhabitants, I = the number of inhabitants in these cities.
'Alpine area' incorporates communal territories classified according to altitudes of communal centers in 1990, see Bätzing 1993, 75. Sources and criteria for the cities as in table 4.1.

In general terms, economic development was uneven in many respects. Although urban agriculture declined, in some places it remained important at the end of the century. In 1901, in Sondrio, a city of 7700 inhabitants in the middle of a wine-producing region, almost half of the active population worked in the agricultural sector, a proportion that was even higher than in the southern French Alps, a traditional stronghold for this phenomenon.[38] On the other hand, even when it was accelerating, industrialization did not have the same effects everywhere. The only city on our sample list whose population decreased substantially between 1870 and 1900 is Glarus, the center of a textile region that had developed much earlier. In many regions though there was growth of previously unrecorded proportions. For example, beginning in the 1840s, glove production in Grenoble entered a new phase of expansion. Between 1866 and 1893, just before the take-off of the modern economy, glove manufacturing employed 25,000 to 32,000 workers in Grenoble and its region. In other places as well, industry attached itself to already-existing forms of production. In Styria the late nineteenth-century expansion of the mining and iron-working industries was so energetic that five new cities appeared in the central Mur valley: Bruck, Leoben, Donawitz, Knittelfeld, and Fohnsdorf.[39]

The structure of the various sectors of public administration (including health and education systems), often favored the large cities that had already enjoyed dominant positions under the Old Regime. Levels of militarization also increased, partly due to the more prominent boundary-marking function of the Alps amidst new nation states, and this could have considerable effects on Alpine cities. In 1891 one quarter of the active population of Grenoble, a city that had long housed an important garrison, were soldiers (8900 in all). In Briançon the military population increased from 460 in 1872 to just under 2700 in 1901. Without these soldiers, the city would not have classified as an urban area according to our parameters.[40] Likewise, those places that expanded over the course of the century as therapeutic and health resorts would otherwise not have qualified for our sample. Even though this sort of tourism was motivated by anti-urban impulses—one went to the mountains to be reinvigorated by nature—and was limited to elites, it had an incipient urbanizing impact. Ischl in the Salzkammergut enjoyed a unique reputation because from mid-century onward the imperial family of Austria had its summertime residence there. But the above-mentioned case of Davos provides an example of expansion that was even more rapid. Here this development was completely tied to the location's altitude and pure air, which were expected to exercise a healing effect on various kinds of pulmonary maladies. In 1870, shortly before the arrival of the first winter tourists seeking out the climate's benefits, Davos numbered 2000 inhabitants. Twenty years later, when the railway was inaugurated, the number had doubled, and by 1900 it had doubled again.[41]

In the late nineteenth century, two cities located respectively at elevations of 1300 and 1600 meters—Briançon and Davos—came into being, representing urban concentrations at higher altitudes than any of the Old Regime cities. As table 4.4 shows, the altitudinal distribution of Alpine cities and populations underwent other notable shifts as well. Contrary to what obtained during the preceding period, the number of cities in the second group (over 500 meters in altitude) increased faster than any other, a fact that points to the diminishing importance of agriculture in the immediate surroundings.

But one should connect this interpretation to other ones. In the first place this change was partly the result of the eastern Alpine regions catching up to what had been happening elsewhere, as has been mentioned above, and was thus not explicable in terms of environmental conditions alone. Second, it remains the case that our representative sample is small and above all irregularly distributed. The rapid growth rate in the middle elevations also corresponds to an initial statistical phase in which small increases enable many places to classify as cities, creating a strong impression. Seen in absolute terms, the difference between low and middle elevations did not decrease in the first period, but increased. It seems that the growth of the nineteenth century was of substantially greater scope, but precisely for this reason it was far more subject to the influence of urban concentrations and their internal dynamics. One finds a particular indication of this when one compares developments in the Alps to those of surrounding areas. At the beginning of this chapter, we approximated that there were about 0.6 cities per 10,000 square kilometers in the Alps in around 1800, compared to 5.0 cities for the same amount of territory in the surrounding areas whose schematic outline we provided. For the period around 1900 a different way of estimating suggests that these densities were 2.1 for the Alps and 12.4 in surrounding areas. Thus over the course of a century, the size of the gap more than doubled.[42]

During the early modern period, by way of summary, urbanization in the Alps was just beginning.[43] Around 1500 probably only one city met our threshold requirement of 5000 inhabitants with a concentrated population of at least 3000 people. Around 1600 there were probably six such cities, in 1700 seven, and in 1800 nine. If we limit our investigation to these major centers and exclude the small cities that are difficult to evaluate, sociopolitical forces emerge as important factors in urban growth. Conversely, economic circumstances were the elements that played a consistent role in delaying growth. As altitude increased, the possibilities of agricultural intensification decreased, cutting into the likelihood of finding sources for urban provisioning. This was also important for low-altitude settlements, because substantial portions of cities' immediate vicinities were typically located in truly mountainous territory. As far as internal portions of the Alps are concerned, urban development seems to have proceeded more rapidly at lower elevations than in high valleys, which were limited by restrictive conditions. In areas surrounding the Alps, urban densities and population levels increased to a far greater degree between 1500 and 1800 than they did in the Alps.

During the nineteenth century these tendencies accelerated to an extent that produced a partial break with the past. Even according to more restrictive parameters, the number of Alpine cities with at least 5000 inhabitants increased from nine to forty-two. At the same time, their altitudinal distribution shifted, such that the highest city was no longer located at 700 meters, but at 1600. But the density of urban centers across Alpine territory continued to differ from what one found in surrounding areas, and this difference increased sharply over the course of the period. With the transportation revolution and other transformative processes, importance shifted from the agricultural hinterland to the internal dynamics of the cities. Rather than resulting in a leveling-out between various regions, this shift caused differences that already existed to become even more prominent.

Chapter Four

The unequal urban growth in the Alps and in surrounding areas helps explain how, between the beginning of the early modern period and the early twentieth century, the Alpine arc came to be a border region between European nation-states and, for centuries, a zone of emigration. Yet the same difference attracted contemporaries in increasing numbers and in an almost magical way, as the records of therapeutic and climatic tourism of the nineteenth century indicate. However one might wish to evaluate these trends, one phenomenon must be clearly articulated. The unequal urbanization of the Alps and the plains cannot be explained in the categorical, ahistorical terms that either anti-urban or anti-rural ideologies are quick to offer. In reality, the processes in both cases were similar, but they worked themselves out under different sets of conditions.

Endnotes

1 *Storia d'Italia*, ed. Giulio Einaudi (Turin, 1973), 5: 367; Lino Colliard, *Vecchia Aosta* (Aosta, 1986), 195.

2 Dematteis 1975, esp. 7–16, 22; for French and German geography see for example Paul and Germaine Veyret, *Au cœur de l'Europe - les Alpes* (Paris, 1967), esp. 300–301, 511–17, 530–31; Günter Glauert, *Die Alpen, eine Einführung in die Landeskunde* (Kiel, 1975), 60–63.

3 For work on urbanization see Rodger 1993; a few countries with Alpine regions are not discussed by Rodger, for these see *Österreichs Städte und Märkte in ihrer Geschichte*, ed. Erich Zöllner (Vienna, 1985); François Walter, *La Suisse urbaine 1750–1950* (Carouge-Geneva, 1994). For a rich overview of the literature and an essay on Alpine urbanization see Gerosa 1988 and Abel Poitrineau, "Villes et campagnes dans les Alpes du XVIIIe au XXe siècle," in Bergier and Guzzi 1992, 175–98. Favier 1993 is an important regional study.

4 It would be stimulating, for example, to compare political culture in the Alpine space with regards to urban classification, as has already been done for a few territories; for Austria see Michael Mitterauer, *Markt und Stadt im Mittelalter. Beiträge zur historischen Zentralitätsforschung* (Stuttgart, 1980), esp. 278–304; he shows the impact of regional structures of dominion on urban status relations, though with a tendency to extend this influence to the overall urbanization process.

5 Bairoch 1988.

6 Expilly 1765 (see appendix to table 4.1), 933; Julien Coste, *Vallis montium. Histoire de la Vallée de Barcelonnette, Hautes terres de Provence, des origines à nos jours* (Gap, 1976), 175; I-Statistics 1871, vol. 1, pt. 1, 398.

7 If one distinguishes between "pre-existing" and "new" cities in different zones and compares their growth levels, one finds that growth levels of "new" cities within the Alps were significantly higher than those of "new cities" in the surrounding areas and also at the foot of the mountains.

8 This excludes the cities at the foot of the mountains. In this analysis the area of the Alps is set at 180,000 km^2 and the total area considered at 578,700 km^2, of which 398,700 km^2 are in surrounding areas. In the Alps, extensive areas are reputedly non-productive, but on the other hand the Alpine territory is artificially shrunk by the flattening out of the surface area; I think

Cities

it advisable to allow these two distortions to cancel each other out (see chap. 2). The differences between Bairoch's data and those in table 4.1 might be an indication that the density inequality was actually greater: in the Alps I have decreased the number of cities by 20%, and at the foot of the Alps by only 8%.

9   The mining center of Schwaz included in table 4.1 might have had a level of demographic concentration that was too low (see n. 17). For the years around 1500, places that were nearing the parameters established here, and perhaps surpassing them, were Gap, Bolzano/Bozen, Trent, and Innsbruck (based on various data that are typically uncertain and inflated).

10   Among the numerous monographs that examine individual cities, the classic works for three large centers should be mentioned here: *Histoire de Grenoble*, ed. Vital Chomel (Toulouse, 1976); Otto Stolz, *Geschichte der Stadt Innsbruck* (Innsbruck, 1959); *Die Landeshauptstadt Klagenfurt. Aus ihrer Vergangenheit und Gegenwart*, ed. Gotbert Moro, 2 vols. (Klagenfurt, 1970).

11   Janin 1968, 493; see also n. 2 above.

12   Daniel Scheitlin, "Contribution à l'étude de la population urbaine suisse: 1200–1850. Constitution d'une banque de données et analyse des résultats," *License* thesis, University of Geneva, 1985, 1: 20–21; Blanchard 1938–1956, 6: 706–11, 741–45; Herbert Hassinger, "Der Verkehr über Brenner und Reschen vom Ende des 13. bis in die zweite Hälfte des 18. Jahrhunderts," in *Festschrift F. Huter. Tiroler Wirtschaftsstudien* 26 (Innsbruck, 1969), 137–94.

13   Mathis 1977, esp. 28, 39, 43, 81, 135–36, 140–43, 154; in Susa, following the improvement of the road but prior to the construction of the railway (ca. 1850), there seem to have been only ten families specializing in the transport trade, see Blanchard 1938–1956, 6: 701.

14   For Grenoble and Klagenfurt, see Favier 1993, 289; Moro 1970 (see n. 10), 1: 405, 420; in the city of Aosta, where the population doubled between 1500 and 1800, the transport sector in 1798–99 was insignificant (Colliard 1986 [see n. 1], 1: 211–13); it was more important in Bolzano/Bozen, which nonetheless had to share its transalpine traffic with Bronzolo/Branzoll (Franz Huter, *Beiträge zur Bevölkerungsgeschichte Bozens im 16.–18. Jahrhundert* [Bolzano/Bozen, 1948], esp. 28–30, 88–89).

15   A correction of older views can already be found in Jean-François Bergier, "Le trafic à travers les Alpes et les liaisons transalpines du haut moyen âge au XVIIe siècle," in *Le Alpi* 1974–1975, 3: 1–72; for a more recent regional study see Jürg Simonett, *Verkehrserneuerung und Verkehrsverlagerung in Graubünden. Die "Untere Strasse" im 19. Jahrhundert* (Chur, 1986).

16   Michael Mitterauer, "Produktionsweise, Siedlungsstruktur und Sozialformen im österreichischen Montanwesen des Mittelalters und der frühen Neuzeit," in *Österreichisches Montanwesen. Produktion, Verteilung, Sozialform*, ed. Michael Mitterauer (Vienna, 1974), 234–315, citation from 239; also Mitterauer 1980 (see n. 4), esp. 302–303.

17   Franz Mathis, "Die wirtschaftliche Entwicklung in der frühen Neuzeit (1519–1740)," in *Tiroler Wirtschaftschronik Nordtirol/Südtirol* (Vienna, 1994), 80–82, 93–96; *Österreichisches Städtebuch*, ed. Alfred Hoffmann, vol. 5, *Tirol* (Vienna, 1980), 211–27, esp. 217 (there were 612 houses in the center in ca. 1780); the figures in table 4.1 are probably high for the first period, and likely low for the half-century between 1750 and 1800.

18   Moro 1970 (see n. 10), esp. 1: 22–36, 102–44, 405–33, and 2: 7–17, 234–39, 292–97 (numbers are provided in 1: 107, 125, 265, 420, and 2: 9); *Handbuch der historischen Stätten, Österreich*, vol. 2, *Alpenländer mit Südtirol*, ed. Franz Huter (Stuttgart, 1978), 204–7, 252–57.

19   Favier 1993, esp. 55–73, 254–301, 403–14, 442–52, citation from 301; Chomel 1976 (see n. 10), 63–245.

Chapter Four

20   Jan de Vries, *European Urbanization 1500–1800* (London, 1984).
21   Paul Bairoch, *De Jéricho à Mexiko. Villes et économie dans l'histoire* (Paris, 1985), see for example 636–39; Boserup 1981, 63–75, 95–97.
22   See chapter 2; a wide-scale description of regional population densities exists only beginning with the early nineteenth century: Helmut Haufe, *Die Bevölkerung Europas. Stadt und Land im 19. und 20. Jahrhundert* (Berlin, 1936), esp. map 1.
23   *Histoire de la France urbaine*, ed. Georges Duby, vol. 3 (Paris, 1981), 57; Fernand Braudel, in *Civilisation matérielle, économie et capitalisme, XVe–XVIIIe siècle* (Paris, 1979), 3: 240, doubles the estimate of Wilhelm Abel, reaching a comparable figure.
24   Hans Bobek, *Innsbruck. Eine Gebirgsstadt, ihr Lebensraum und ihre Erscheinung* (Stuttgart, 1928), 34, 86–87; for the link between altitude and agrarian intensification see chapter 3.
25   Calculations made on 1:200,000 topographic maps; percentages of the areas of the outskirts (within a 15-km radius) of Alpine cities under 1000 meters, whose numbers are not provided in the text, are: Gap 38, Grenoble 67, Sisteron 75, Belluno 72, Bolzano/Bozen 47, Rovereto 58, Trent 70, Klagenfurt 97, Schwaz 32; the median for Alpine cities is 53.
26   The number of inhabitants for Belluno and Gap in table 4.1 is high; see Ferruccio Vendramini, *La rivolta dei contadini bellunesi nel 1800* (Feltre, 1972), 30 and Eugenio Sief, "Il comune di Belluno in età napoleonica (1805–1813)," *Laurea* thesis, Catholic University of Milan, 1990–91, 234; Paul Guillaume, "Mouvement de la population du département des Hautes-Alpes au XIXe siècle," *Bulletin de la Société d'Etudes des Hautes-Alpes* (1908): 210. The two other places without river traffic were Aosta and Susa, see Brönnimann 1997.
27   Vendramini 1979, 85, also 31–32, 87–93; Vendramini 1972 (see n. 26), 30.
28   Favier 1993, 186–91, 263–65; for the high percentage of agricultural workers in the southern French cities, based on data from successive periods, see Bernard Barbier, *Villes et centres des Alpes du Sud. Étude de réseau urbain* (Gap, 1969), for example 21–24, 33, 38; important evidence concerning medieval levels of development is provided by the extraordinary density of diocesan centers, see *Atlas zur Kirchengeschichte. Die christlichen Kirchen in Geschichte und Gegenwart*, ed. Jochen Martin (Freiburg im Breisgau, 1987), 71.
29   Favier 1993, esp. 188–92, 263, 421, 469, citation from 188.
30   Bätzing 1993, 75; his definition of the Alps is a bit broader than that of Dematteis, which has been employed in this chapter; communal structure was modified during the nineteenth and twentieth centuries, to the advantage of communes located at lower altitudes.
31   The average percentage of area under 1000 meters within a 15-km radius of low-lying cities (under 500 meters) is 72%, for higher cities (500 meters and higher) it is 31%.
32   Favier 1993 (see n. 3), esp. 146, 157–59, 165, 175, 179–82, 363–65; similar examples of provisioning problems were experienced by mining centers, see for example Mathis 1994 (see n. 17), 91, 95, 105; *Bad Ischl. Ein Heimatbuch*, ed. Ischler Heimatverein (Linz, 1966), 164–66, 200, 236, 395.
33   In Belluno the communal territory was sharply expanded during the Napoleonic period, artificially increasing the size of the urban population (16,000 inhabitants for 1871), see Sief 1990–91 (see n. 26); the formation of urban concentrations that extend beyond the commune's territory can be partially examined through the birth of new cities (or suburbs), see below for this phenomenon in the case of Innsbruck.
34   Since only certain years around 1870 and 1900 have been taken into account, a few cities that satisfied the criteria for short periods between these dates (such as Borgo in the Trentino) could be

Cities

overlooked. If the additional criteria used in table 4.1 were imposed, the number of the Alpine cities would reach 29 (for 1870) and 54 (for 1900); see the appendix, table A.6.

35   See chapter 2 above.

36   Problems related to geographic classification make urbanization percentages (urban population compared to total population) a fairly arbitrary criterion in contexts like this one, and for this reason we will not attempt to provide corresponding data; for France see the contemporary description by Émile Levasseur, *La population française. Histoire de la population avant 1789 et démographie de la France comparée à celle des autres nations au XIXe siècle*, 3 vols. (Paris, 1889–1892), 1: 326–29, 3: 304–31; for the flood of immigration into Grenoble see Chomel 1976 (see n. 10), 250, 288.

37   Pierre Brunner, *Les chemins de fer aux prises avec la nature alpestre* (Grenoble, 1935), 8–38; the railway had a strong impact on Innsbruck, where it helped create a concentration of regional traffic (Bobek 1928 [see n. 24], esp. 44, 48, 55–67).

38   I-Statistics 1901, vol. 3, 33–81 (active population without counting retired persons and family members); for southern France see above, n. 28.

39   Chomel 1976 (see n. 10), 256–60, 291, 308–9; Zöllner 1985 (see n. 3), 83, 97–98, 106–7.

40   Chomel 1976 (see n. 10), 290 (active population without counting retired persons and family members); F-Statistics 1872, 225; F-Statistics 1901, 93 (population counted separately); Paul and Germaine Veyret, "Petites et Moyennes Villes des Alpes," *Revue de Géographie Alpine* 52 (1964): 105–7.

41   Both places had decentralized residential structures, such that Ischl did not satisfy (not even for 1900) the concentration parameter adopted here, even though the commune numbered 9700 inhabitants. For the "imperial" spa see Bad Ischl 1966 (see n. 33), 232–35, 642–50; for Davos, *Inventar der neueren Schweizer Architektur 1850–1920*, ed. Gesellschaft für Schweizerische Kunstgeschichte, vol. 3 (Berne, 1982), esp. 317–78.

42   Calculations based on *Andrees Allgemeiner Handatlas*, 5th ed. (Bielefeld and Leipzig, 1913), which lists cities according to population sizes in around 1900. In surrounding areas (as defined above) I counted 496 cities, and in the Alps 38 (the zone at the foot of the Alps was once more excluded). Comparison with figures collected by other methods shows that these density figures are too low for 1900, while those for 1800 are too high.

43   Already during the favorable demographic conjuncture between the early and late Middle Ages it seems that some places displayed quite significant concentrations. Bairoch points to 6000 inhabitants for Sisteron in 1300, and 8000 for Gap in 1400; the southern French Alps were thought to have a particularly large population during this period (Bairoch 1988, 26, 29).

5 Environment and Development

"Alps: *Alpes* is the term used by Strabo, Ptolemy, Herodianus, Pliny, Caesar, and other Greek and Latin writers to refer to the high mountain chain of 188 miles in length that separates Italy from Germany, France, and the Swiss confederacy." Thus was informed the elite circle of readers of the first volume of the *Helvetisches Lexikon*, published in 1747. The individual sections of these high mountains, continued the entry, are known by various names and have been discussed in particular by Josias Simler. In his treatise, the learned scholar from Zurich described in detail the "names, length, width, and height" of the Alps, and referred to "Hannibal's first crossing of them, and to other difficult crossings, undertaken in our days, and to peoples, bodies of water, trees, plants, and animals."[1]

In certain ways the Alpine space is among those areas in which the relationship between the environment and development is especially open to observation—a theme whose importance extends to many disciplines at present. On the one hand, that to which we typically refer as 'the environment' makes its flashy appearance in the world of the mountains, with their "bodies of water, trees, plants, and animals." On the other hand the Alps are situated at the intersection of countries animated

Environment and Development

by lively research activities. Already during the period that concerns us here, many serious and methodological studies about the mountains had been written. Some of these became well-known, and continued to be discussed for quite a while, such as Simler's *De Alpibus Commentarius* (1574), or the ideas of Friedrich Ratzel on *Die Alpen inmitten der geschichtlichen Bewegungen* (1896) (The Alps in the Midst of Historical Movements). Of course, the contexts in which scholars from the sixteenth through the nineteenth centuries operated were very different from the conditions of present-day research. For example, the ancient canon, "the Greek and Latin writers," were known and appreciated in ways that are difficult to understand today. But the strangeness of natural phenomena and human settlements located only a short distance away from their own cultural centers aroused in many a kind of personal and empirical curiosity. The development and specialization of scholarly research that took place in the nineteenth and (above all) twentieth centuries led to rapid changes in the size of available bodies of knowledge, and in the models that structured this knowledge. Naturalists and geographers tell us that today the Alps are "the best researched mountain space in the world."[2]

From an historical perspective, however, one must acknowledge that it is still difficult to construct a comprehensive picture of the demographic and economic development of this area, even if one only begins at the end of the Middle Ages. Such a picture is important, though, since the historical discipline is particularly committed to the reconstruction and interpretation of how the conditions of human society changed over time. The preceding chapters have discussed the history of population levels, of agriculture, and of cities in the Alpine space. Now it is time to pause, assess, and take into account the ways in which these different sectors related to one another. Next we will extend this clarification to relations between the Alps and surrounding areas. Then the theoretical implications of these findings will be discussed. This chapter once again confronts us with the limitations of the available historical records, but it also shows that the theme of environment and development lends itself to examination from multiple perspectives. We need to ask ourselves which perspective makes most sense in each given context.

An intermediate assessment: differentiated growth

In order to focus on the importance of environmental factors, we must first keep in mind the long-term tendencies that have already been identified. From 1500 to 1900 the Alps underwent a process of demographic, agricultural, and urban expansion (1);

the growth that occurred after 1700 differed from that of the previous two centuries in that it was often slower in the higher elevations than in low-altitude regions (2) and slower in the Alps in general than in surrounding areas (3). The importance of these general tendencies should not be overestimated, since in each period one could find areas that developed differently. Trends such as these become apparent when observing a vast area over the long term, which is quite a challenge given the fragmentary nature of the documentation and the actual state of the research. This has resulted in quantitative data most of which is based on approximate estimates.[3]

(1) Between 1500 and 1900, the population of the Alps appears almost to have tripled. Within the Alpine space as defined by Werner Bätzing, the absolute numbers at hundred-year intervals were 2.9 million in 1500, then 4.0, 4.4, 5.3, and 7.9. The population increase and the high percentages of agricultural workers that can be documented for many places indicate that during our study period there was substantial growth in agricultural production. Increased cropping frequency per surface area, both for forage and for plant cultivation, was an important dimension of agrarian intensification. But the status of the available documentation makes it easier to document changes in the livestock population and in the variety of plants that were cultivated. The shift from sheep-raising to cattle-raising and the introduction of corn and potatoes figured prominently in this process. New, more intensive forms of environmental exploitation generally required consistently more labor. Thus, there were a number of ways to intensify agricultural production that were only initiated once demographic pressure simultaneously increased demand for foodstuffs and expanded the available labor force. For example, potato cultivation spread throughout the Alpine space at the end of the eighteenth and especially in the early nineteenth century, but it had been known in all of these countries for two hundred years. If one considers that potato farming not only increased output per surface area, but also required a heavy labor investment, it is easier to understand why it took so long for this crop to catch on.

Our investigation of urban development relied on Giuseppe Dematteis' definition of the Alps and on a threshold of 5000 inhabitants for a place to qualify as a city. According to these parameters, the Alpine space probably only had one city in 1500, around nine in about 1800, and forty-two at the turn of the twentieth century. Urbanization was linked in various ways to demographic expansion and to increased agrarian production, though in the formation of the large cities sociopolitical factors were centrally important. The history of Grenoble, by far the largest city in the Alps since the seventeenth century, was for example marked by a progressive concentration

Environment and Development

of political power, in a process that was initiated by the provincial nobility and then increased under the growing centralization of the French state.

(2) We have defined as Alpine regions all of those modern, politically consolidated administrative units of a certain size, at least 75% of whose territories are located in the Alps. If these regions are classified according to the average altitude of their communal centers, one can analyze general demographic developments, at least provisionally, with reference to factors linked specifically to elevation. In so doing, one is confronted with more and more phenomena during the eighteenth and nineteenth centuries that slowed the growth rates in high-altitude regions compared to low-altitude ones, while in the earlier period demographic changes in these areas were more equally balanced. A number of indicators also suggest that, with respect to Alpine agriculture during the later period in question, the difference in levels of intensification at different altitudes grew particularly rapidly. Thus, during the sixteenth and seventeenth centuries, large portions of low-lying river valleys continued to be used for pasture or gathering wild fruits. When small-scale water control projects began to be carried out, there were measurable production increases in many places. Then, in the eighteenth and nineteenth centuries, river courses began to be corrected and large-scale drainage and irrigation structures were created, turning bottomland into the preferred locations of agrarian intensification.

According to our evidence, in around 1600 there were six cities of at least 5000 inhabitants in the Alpine space. In the period that followed, in lower altitude areas, the urban centers grew larger, slowly at first and then rapidly, such that by 1800 the urban population had doubled, while the number of cities at higher altitudes did not change. During the nineteenth century the growth process was different—the number of low-altitude cities grew less quickly, and high-altitude cities began to appear. By about 1900 Davos, located at almost 1600 meters, exceeded by far the altitude of what used to be the highest Alpine city. The economic dynamics that underlay urbanization processes began to shift rapidly: while early modern cities had been closely dependent on their agricultural hinterlands, the transportation revolution of the nineteenth century created ample new opportunities for urban growth.

(3) Efforts to compare the Alps with surrounding areas run up against the problem that the concept of "surrounding areas" lacks precise territorial definition. It seemed reasonable to me to make reference to such areas in different ways depending on the particular problem at hand and the availability of data. According to existing estimates, the demographic trends in the Alpine space developed in the same direction as in the

rest of the present-day countries in which the Alps are located. Population increases in the two zones differed only slightly during the earlier period, and then after 1700 one can identify a constantly growing disparity between slow rates of increase in the Alps and faster rates in the entireties of the countries concerned. Comparison of large regional areas in the Alps and in surrounding territories provides similar results. During the sixteenth and seventeenth centuries there were some regions in which the growth rates in mountain zones were higher than in nearby lowlands. Later, the increase in each of the Alpine regions examined fell behind that in the surrounding regions, and at the same time the differential between growth in these areas expanded in absolute terms. In the agricultural sector progressive intensification increased the importance of the length of the growing season, thus rendering altitude more significant as well. As fields were exploited more frequently, the time factor became more critical. A good example is provided by the different intensity levels in cropping frequencies for hay. During the early modern period, in the Lombard plain and in low-lying river valleys, some places began to be mowed four times per year, and in the plain of the Po, forage grasses were cut up to seven or eight times. At higher altitudes with shorter growing seasons, grass grew so slowly that in many cases it was cut only once every two or three years.

In absolute terms, urban density also increased far more in surrounding areas than in the Alpine space. This was particularly pronounced during the eighteenth and nineteenth centuries. Most Alpine cities were located at relatively low altitudes (until 1800 their average elevation was only 460 meters), but suburban areas almost always included a significant percentage of higher-altitude terrain, making urban growth difficult from the perspective of food supply. During the nineteenth century, when the agricultural hinterland became less important, the differences between the Alps and surrounding areas grew even more. Cities then could expand at higher rates, but this expansion was conditioned more than before by the internal dynamics of those urban centers that had already existed.

On the whole one can conclude that environmental factors played a variable role in historical developments. Population growth and agricultural intensification meant that the impact of elevation became more critical, and that diversification both inside and outside of the Alpine space increased. But at the same time, the influence of geographic conditions could be attenuated, as shown by the transportation revolution and nineteenth-century urbanization. Close attention should be given to the following perspective: even when differences between growth rates were minimal, the density

Environment and Development

differential between the Alps and surrounding areas increased notably, because beginning in the late medieval period and even earlier the mountains were less populated and urbanized than surrounding regions, and thus had a different point of departure for the growth of the early modern period. Obvious proof of this fact is provided by the unequal distribution of cities whose population numbered at least 5000 in around 1500 (see map 4.1). As the difference in growth rates increased, as became the norm after 1700, the overall population disparity became significantly greater.

Many of the relations between the Alps and surrounding areas can be understood with respect to this density differential, and some of these relations even help to explain the differential. In what follows I will mention lumber transport and livestock trading, for which the density factor is particularly noticeable. Then issues related to migration and state integration will be discussed. When contemporaries distinguished between the mountains and the plains, they often did so with reference to the local or regional situations that were immediately apparent to them. In order to avoid misunderstandings or unfounded representations, it is stressed that our general reflections on the relations between the Alps and surrounding areas is based on a modern perspective.

Relations between the Alps and surrounding areas

In some of the years around 1560, between the small Tyrolean community of Stanzach and the southern German city of Augsburg, from 250,000 to 300,000 logs were floated down the Lech River annually. This amounts to a daily average of up to 5,000 logs. On the basis of toll registers that were compiled and preserved in various transit points, one can gain a detailed insight into this lively traffic in timber for the early seventeenth century. The sources show that this traffic occupied all of the affluents on the south shore of the Danube, and that the timber was sometimes destined toward quite distant points of sale. In 1603 an author from the south side of the Alps described a similar situation there. He reported that the lateral valleys of the upper Ticino basin were extraordinarily rich in woodland, including timber that could be used for ship masts and galleys, and especially for beams used in house construction. This timber was all transported by water, toward the Lago Maggiore, "whence it was then sent to Milan, Pavia, Cremona, Venice, and throughout Italy."[4]

The timber trade, or rather timber transfer (sometimes the wood included fiscal contributions or timber harvested from state forests) is among those phenomena that characterize the long-term history of the Alps. The basic elements of this transportation structure are widely documented, and have already been referred to above. On

Chapter Five

one side there were a great number of wooded areas, and on the other a high population density that had thinned out the timber resources in wide areas, creating a strong demand—especially in large cities—for wood to be used in a variety of ways. The waterways that flowed from the Alpine arc toward surrounding areas offered a means of transporting heavy and voluminous goods relatively inexpensively.[5]

Demographic and economic expansion caused this demand to increase in the eighteenth century and then peak in the nineteenth, before wood lost its importance as a source of energy with the advent of fossil fuels. On the technical side, the development of transportation methods is an indicator of the long-term conjuncture. In contrast to what took place far downstream, where logs were tied together to form rafts onto which other merchandise was then loaded, on upriver portions the logs were usually floated singly, carried along by the current. In addition to this ancient method, other costlier methods were also employed, making it possible to exploit Alpine forests even further. Especially from the eighteenth century on, well-built timber slides, made of wood and multiple kilometers long, appeared in many side valleys. These were equipped with blocking valves so that when the water flow was too low, water could be collected and then released in a full wave. An amazing example of investment in transport is the complex of canals, some carved through mountain tunnels, constructed from 1822 onward at the Preintaler Gscheidl, north of Semmering in Austria. The rapid growth of Vienna and the consequent sharp increase in the demand for wood had motivated the construction two decades earlier of the Wiener Neustädter Canal at the base of the Alps. The canals that were built at higher elevations on the other side of the Preintal, part of which could be used not only for floating timber but also for navigation, were laid out in the midst of mountainous terrain. Wood was carried along through tunnels, on a rather long itinerary that passed beneath the watershed of the mountains (see illustration 10).[6]

If the "wood shortage" of the eighteenth century generated widespread public debate among contemporaries, in the nineteenth century "the protection of mountain" forests became a hot political topic, resulting in state intervention in all of the Alpine countries. This development was stimulated both by the changing economic environment and by the growth in centralized forest administrations, which were linked generally to the consolidation of nation-state institutions. Central to these discussions was the problem of floods and threats of flooding in the valleys, which were attributed to deforestation in mountain areas. State protectionism emerged rapidly and vigorously in the French Alps, where methods of exploiting the woodlands had been stigmatized by the publication of *Etude sur les torrents des Hautes-Alpes* (Study of the Mountain Streams of the High Alps) in 1841. "The work provided justification for taking forest management away from peasants," writes the anthropologist Harriet G. Rosenberg, "by apparently demonstrating that communal practices of pasturing sheep in forests, as well as extravagant woodcutting and clearing, had drastically increased river flooding." Already a century earlier there had been efforts by the center to intervene in this area, but it was the French forest law of 1827 that laid the modern juridical foundation, providing state offices with strong support in the debates that would not die down. According to Rosenberg, these legal provisions had a considerable impact on the local economy, creating problems that were then used to justify subsequent interventions. In other regions this institutional infrastruture was often missing, though the desire to impose similar kinds of protectionist policy was not. In Switzerland, which had a completely different kind of state structure, the first national forest law of 1876 only addressed "the high mountains."[7]

The impact of population density is also prominent in the livestock trade between the Alps and surrounding areas. To begin to frame this problem on a macro-level, one can consider a map of the long-range European trade in large livestock, based on the studies of Othmar Pickl and Wilhelm Abel and focussed (in a fairly approximative yet coherent way) on the situation in around 1600. This map shows that the most important zones of bovine raising and exportation during this period were in the lightly populated "external pasture areas" of eastern Europe. Thus, for example, from the low plains of Hungary and the principalities of Moldavia and Walachia, about 200,000 heads of cattle were driven to slaughter each year in more densely populated and urbanized centers of consumption in the West, sometimes at distances of 1,000 to 1,500 kilometers. Demand for meat was especially high in medium-sized and large cities. Already in around 1500, Venice acquired about 15,000 heads of beef from Hungary each year, and by about 1600 these imports increased by 50% (the population of Venice had increased in the meantime from about 100,000 inhabitants to 151,000). This commercial traffic from the East also extended into the Alpine regions. For example, part of the livestock that ended up with butchers in Innsbruck came from Hungary. On the whole, though, the Alpine space was without a doubt a zone of exportation: the map mentioned above shows that this was the only specific zone of large livestock-raising in the middle of the continent, with an annual export of about 50,000 heads.[8]

Alpine animal husbandry differed from that carried out in the low plains of eastern Europe in that pasture activities, which required a great deal of space but little labor, were limited to the warm period of the year. However, markets were closer and transport costs lower. The long-range cattle drives from eastern Europe reached a peak in around 1600 and then diminished over the long term, because the process of agrarian intensification in the eastern points of origin led to steady increases in grain cultivation there, while in the West livestock-raising expanded. Available studies indicate that livestock exports from the Alps, originating in contexts of markedly different densities and intensities within a relatively small area, maintained constant volumes, if not increases, through the nineteenth century.[9] We can safely assume that for centuries the main markets for these exports were in northern Italy, an area that was much more urbanized, especially in the sixteenth centuries, than other nearby regions. Urban and industrial growth usually resulted in increased demand, in some instances to such an extent that the prices for animal products surpassed those for cereals, even as overall demand continued to expand.[10]

Most of the exports from Carinthia, which were often in the hands of privileged monopolies during the seventeenth and eighteenth centuries, were directed toward the Venetian lands, and amounted to about 1000–2000 heads per year. In 1635 it was said, with reference to the Savoyard district of the Beaufortin, that being there was like having reached the ends of the earth. Still, most people there were well-off "because of the great quantity of horned animals that they feed, and that they regularly send to Piedmont." In the Swiss area, almost the entire mountainous region was oriented toward the *Welschlandhandel* (commerce with Italy), that was particularly intense in some of the territories of the northern slope of the Alps. In September 1527, following a controversy concerning politically-imposed commercial restrictions, the inner Swiss cantons underscored their habitual practice of raising livestock over the course of the entire year and then either bringing it to Milan or selling it to merchants who came from there.[11] In 1513 the Swiss confederates had permitted the creation of a market in Lugano to be held during the second week of October. It became the most important emporium in that part of the Alps, and in the eighteenth century it was calculated that several thousand heads of large animals were transported there. The customs records of animals in transit over the St. Gothard pass are more accurate—they show that for several years around 1725 there was an annual average of 3210 heads, increasing to 5810 in 1806 and 7690 in 1832.[12]

Swiss bovines were used in Lombardy primarily for the dairy industry and only secondarily for slaughter. According to first-hand accounts and to the work of economists, intensive cultivation of the countryside made it disadvantageous to raise livestock in Lombardy itself. When one takes into account the profits earned from cheese production, which increased sharply in this region during the eighteenth century, it was considered preferable to import livestock, despite higher prices. This trend toward dairy production affected the large-animal trade in very different ways, both from region to region and over time. It appears that in some Alpine regions that had increasingly oriented themselves to cheese exportation over the course of the Old Regime, bovine exports leveled off or even fell.[13]

A notable shift occurred in the nineteenth century when, following the processes of urbanization and industrialization, intensive livestock-raising and dairy production began to expand in almost all of the areas surrounding the Alps and the number of animals in the plains began to increase faster there than in the mountains—even faster than in places such as Styria, where growth was notable. This made it frequently possible to sell bovines at less-distant markets, while cheese commerce took on a new dimension in both lowland and upland zones. Whereas the costs of moving livestock long distances remained relatively low up to the nineteenth century, the revolutionary changes in the transport system favored the long-distance export of dairy products.[14]

Among all of the kinds of relations between the Alps and surrounding areas, migratory movements have always enjoyed the most attention from the scholarly community. On this theme there is a vast and continually-growing literature that continues to renew the field. Within general historical studies of migration, mountains and the Alps in particular are prominent, because they are seen as providing a paradigmatic example of early emigrations. In his book on the Mediterranean area during the sixteenth century, Fernand Braudel even wrote about a "reservoir of men for other people's use," thus coining a phrase that is often cited by those who adhere to the doctrine that mountain emigration results from the overpopulation of areas whose resources are scarce.[15] Modern research has criticized and qualified this approach from a number of perspectives.

For example, in different kinds of studies on migratory movements (mainly temporary and commercial) in the Alpine regions of France and other countries, Laurence Fontaine has emphasized the social logic and internal dynamics of emigration processes. She shows that early modern mobility can be understood as a way of living autonomously in spaces that constantly shifted between cities and countryside. Relations with mountain villages were undoubtedly important, but were of a different character than those represented above all by the French geographers. The objections of Pier Paolo Viazzo to earlier scholarship are also particularly significant: he cast doubt on the traditional overpopulation thesis thanks to original research carried out in Piedmont and

Chapter Five

to a close reading of the literature dealing with many parts of the Alps. According to his interpretation, migratory movement from the mountains varied extraordinarily in both size and form; emigrants were not typically poor, but often belonged to wealthier social groups; beyond the emigration stream, there was also a less widespread immigration movement; migration was not only the consequence of internal demographic shifts, but could also serve as an agent of population changes, via nuptiality. Research has rarely paid close attention to these important and complex phenomena, such that one can appreciate the pessimistic conclusion reached at the end of this author's study. Nonetheless, one must take into account the difficulty involved in collecting evidence about the history of migratory practices, let alone in quantifying these data, which often lend themselves to varied interpretations. Furthermore, historiography should not permit issues of complexity to obscure the utility of generalizations.[16]

In contrast to what has been demonstrated for timber transport and livestock trading, the relationship between population density and migratory processes is ambivalent. Density growth in a given region could be a reason for people to emigrate from there to other regions with farmland that was either more abundant or in a more easily exploitable environment. But density growth could also accelerate agrarian intensification, market formation, and urbanization, thus attracting immigration. The overpopulation thesis only considers the first possibility: roughly from the beginning of the early modern period, the Alpine population was thought to have reached levels in many places that outstripped available resources, resulting inevitably in a century-long emigratory movement. Our findings basically point in the opposite direction. If the population of the Alpine space was able to triple between 1500 and 1900, even though agriculture continued to play an important role at the end of the period, it is difficult to understand why scarce resources around 1500 would have been a decisive cause of growing emigration. A general fact that is often overlooked is that the areas surrounding the Alps were often more populated and above all more urbanized. These basic conditions, together with many other indicators, underscore the significance of pull factors and attractivity for migration processes.

Let us again take Venice as an example. When in 1630 the city on the lagoon lost a great many inhabitants because of the plague, within the space of twelve years—despite a simultaneous decline in rural population pressure—there was a net migration gain of 23,000 people, and by 1655 the arrival of 39,000 immigrants caused the urban population to rise to 158,000 inhabitants. Among these were people from the Grisons, where at the beginning of the century the population had probably been 14 persons per square kilometer. For quite a while already Venice was one of the destinations for temporary emigrants from the Grisons, and later it became their main base of support. When the Serenissima denounced its treaty of alliance with the Grisons in 1765, a whole colony of Rhaetian artisans and merchants had to look for new places in which to carry out their activities. They were able to make use of extended networks of relations that their countrymen had already begun to develop and that were continuing to expand rapidly. In the years before 1900 no less than 594 European cities had immigrants from the Grisons, many of whom had begun to specialize in pastry-making and running modern coffeehouses. Given such developments there is little doubt that this migratory movement was strongly influenced by urban growth.[17]

I believe that we can also situate some chronological and regional variations in migratory flows into a similar interpretive framework. The takeoff of the urban sector beginning in the eighteenth century is quite likely an important reason for the growth disparities between the Alps and surrounding areas—disparities that were generally increasing. The precise percentages of people who emigrated from the Alps are definitely difficult to identify for specific cases, but there are still many indications that during this period emigration assumed new dimensions and could have the effect of markedly holding back population increases in mountain areas.[18]

Available documentation shows that the northeastern parts of the Alpine arc were the regions in which emigration was relatively limited prior to the nineteenth century, while on the southern slopes emigration seems to have been heavy already during the early modern period. During those centuries, from Piedmont to Lombardy and as far as the Veneto, there were some valleys in which the emigration of men, mostly from the artisanal sector, was so strong that it created a complete gender imbalance. In these "vallées des femmes" (Dionigi Albera), agricultural work fell for the most part to the women, and on occasion the scarcity of men even led to the reversal of political roles, such as in the case of the woman who in 1722 became mayoress of the community of Valsassina, in the district of Como.[19] The difference in emigration intensity between the northeastern Alps and the southern slope corresponds on the whole—and in my judgment not coincidentally—to the differences in urbanization that one observes in the Alps and in surrounding areas between 1500 and 1800.[20]

The urban perspective also helps shed light on the particularities of mobility issues in the Alps. To what degree is one justified in attributing to the mountains a particular history with respect to this problem? The average volumes of mobility, that is, the amounts that seem normal for many areas of the plains, are understood in different ways in the historical scholarship on the early modern period, not least because of methodological and other differences for each of the countries concerned. A generally held view, though, is that urban immigrants came largely from immediately surrounding areas, the size of the catchment zone varying with the importance of the city. Many of the people who married or died in the city of Vienne, for example, which was located in the Rhône valley south of Lyon, had been born outside of the city. In the eighteenth century, out of all of the non-Viennois spouses, 43% of the men and 60% of the women came from places no farther than 20 kilometers away. Vienne's 8,000 to 10,000 inhabitants made it at that period ten times smaller than the nearby metropolis of Lyon, where a good half of the immigrants who then married had previously resided in localities up to 50 kilometers away. This sort of mobility model is widespread, though it can obviously take different forms in given cases. But when we examine it in relation to urbanization in the Alps and surrounding areas, it becomes clear that the volume of Alpine mobility should not be exaggerated. Urban "factories," which brought other people under their spell, were mainly situated in the lowlands.[21]

What made mobility in mountain areas distinctive seems to have been not so much the numbers involved, but the high percentage of people who emigrated to far away places and the above-average level of their professional specialization. These two trends were frequently related. In a context of modest population density, productive activities and specific commercial sectors depended on far-flung markets. If the inhabitants of certain valleys wanted to pursue work in a certain field once they discovered it and were attracted by it—and there were good reasons for doing so—they had to configure their activities very broadly in terms of their spatial reach. In cities far removed from their places of origin, these groups stood out both because of their professional specialization and because of the cultural distance between homeland and host city. This visibility helped create the impression in urban centers (and eventually in historiography) that Alpine mobility was singular, even though the levels of mobility in the immediately surrounding areas were quite considerable.[22]

In general terms, then, one can connect very different kinds of relationships between the Alps and surrounding areas to differentials in population density between the two zones. But the general nature of this perspective also limits its utility; this becomes most clear, perhaps, when one looks at processes of state integration.

Sketches of how territories were structured politically have already been provided (chapter 1), and we know that the centers of gravity of states and larger power configurations were almost all located in the plains, either nearby or far away. Many mountain

Environment and Development

areas were typically far removed from authorities and characterized by high degrees of local or regional autonomy. With the intensification of administrative activities and the growing nationalism of the eighteenth and even more the nineteenth century, the Alpine space became more closely tied to various states and their centers. In the course of this process of integration, the importance of certain boundaries—national ones—increased, while that of others—local and regional ones—was weakened. From one perspective, this process appeared to carve up the mountains into subdivisions, but from another perspective, it seemed to open up the Alpine space. Thus, in general terms, the field of political-state activity was undergoing changes that paralleled the location and increase of demographic and economic density described above. Further, political integration itself also contributed to increased differentiation, since the exposure of mountain zones to the new state structures could serve to amplify Alpine migration. Yet the specific ways in which states were formed and power was exercised cannot be fruitfully analyzed with this approach. These issues were obviously important for political life, and they varied widely in different cultural conditions. To understand them we need to examine other circumstances—some of which we will discuss in the next chapters.[23]

History and ecological models

Finally, we must confront, on a theoretical level, the question of how to evaluate the relationship between environment and development in our study area. It is well known that existing studies attribute a great deal of influence to natural conditions in the Alps. Certainly, geographers and anthropologists have long developed areas of expertise related to their specialized interests in the mountains. But references to the environment also appear in many other kinds of studies, including historical ones. Indeed, it is good for research to engage in continual and generalized reflection on themes that cut across disciplinary boundaries. Still, this sort of reflection becomes difficult when the models defined as 'ecological' (because they deal with the natural environment and its use by humans) differ so widely in kind. For this reason it makes sense to start by looking at individual approaches, and to indicate three specific studies that treat differently sized portions of the Alps, and different areas, from the perspective of different disciplines.[24]

The work of Raoul Blanchard, *Les Alpes occidentales*, which appeared between 1938 and 1956 in many volumes and has since become a classic, is a geographic description of the French Alpine areas and of southern Piedmont. The author takes historical sources into account more in the later volumes, especially demographic sources dealing with the eighteenth and nineteenth centuries. In so doing, at several points he expresses his view that the mountain regions suffered from chronic overpopulation. A kind of distillation of this encyclopedic and multi-faceted work on the western Alps appeared in a brief article published in 1952, which summarized Blanchard's view of the life of mountain people as follows: "For the mountaineer, the plains are always the good country, while the reverse is never true." This conclusion is the result of a body of reflection based on environmental variables. Alpine topography forces local inhabitants to engage in back-breaking work in order to derive some benefit from steep slopes, which are further subject to erosion and damaged soil structure. At high altitudes the climate is harsh, with short growing periods and long winters, resulting in large seasonal variations in the amount of agricultural work. Because of their geographic isolation, mountain peasants are left to fend for themselves, following the traditional practices developed by their forebears in order to escape the trap laid for them by their "evil stepmother Nature."[25]

In 1981, with the programmatic title *Balancing on an Alp. Ecological Change and Continuity in a Swiss Mountain Community*, the anthropologist Robert McC. Netting published a widely-discussed study of the Valaisan community of Törbel. He examines above all the issue of how the population of this high-altitude locality was able to survive for so long by living on its own, limited resources. On the basis of a demographic analysis of sources beginning in the seventeenth century, the author reaches the conclusion that internal regulatory mechanisms permitted the community to adapt successfully to its environment, by reaching a homeostatic equilibrium. Soil erosion was as minimally problematic as overuse of the forest. The absence of dramatic famines and of progressive impoverishment suggests that the local population did not outgrow its own resources. "Rather than being a society that periodically exceeded its carrying capacity, only to be ruthlessly cut back, Törbel seems to have approached a homeostatic condition in which density-dependent mechanisms such as a high age at marriage, celibacy, and migration kept population growth within supportable limits." According to Netting, emigration was not prominent in this region. He emphasizes the degree to which the community was closed to the outside, such that it could be considered an "island in the sky." The forces of change, however, did not come from inside but from outside the community: it was the spread of potato cultivation that expanded the economic base and gave rise to population growth.[26]

Environment and Development

A different territorial and thematic context is examined in Michael Mitterauer's historical study, *Formen ländlicher Familienwirtschaft. Historische Ökotypen und familiale Arbeitsorganisation im österreichischen Raum* (Forms of the Rural Family Economy. Historical Ecotypes and the Familial Organization of Work in the Austrian Space, 1986, abridged versions in 1990 and 1992). His model refers to regional and supraregional units that were linked with each other via exchange processes: "According to this perspective, regional economies were not merely the results of a specific mode of adaptation to natural environmental conditions. To an even larger degree, the organization of work was structured supraregionally, in ways that were reflected by local forms of production." With this observation, the author underscores how many different possibilities there were for adapting to the environment, and opposes himself to the dangers created by a static perspective. Mitterauer uses this multi-layered view of ecotypes to make sense of the forms of household and family organization that he reconstructs on the basis of census documents from the seventeenth through the nineteenth centuries. As one might expect given variations in climate and soil properties, historical economic zones within the territory of present-day Austria differed greatly from each other. This heterogeneity extended to mountain areas, although the basic concern of mountain peasants, in contrast to what one found in the pre-Alps and in the plains, was livestock-raising. The continuous nature of the work required by livestock-raising favored the stable employment of agricultural workers, creating an ample demand for a labor force. A far-reaching labor market met this demand, leading to the emergence of a *Gesindegesellschaft*, an ideal type referring to a society of domestic agricultural laborers. By analyzing the organization of work in this way, according to Mitterauer, an environmental perspective can thus helpfully explain the various historic family forms.[27]

These three examples show how both the starting point and the explanatory models used affect the results of the research endeavor. Blanchard transposed Alpine environmental variables directly into the human realm, creating a very unrealistic picture of the mountain world as a generally disadvantaged space in which the same, unchanging adaptive practices were always necessary. Netting and Mitterauer relativize and mediate the link between environment and human society, placing at the center of their investigations (respectively) demography and family, and appealing to environmental factors in order to explain real, observable phenomena. Netting focuses on the endogenous growth potential within a local system, stressing the ability of demographic variables to adapt flexibly to available resources. According to Mitterauer, the forms of local production also express regional structures. Though he is particularly skeptical about the possibility of a direct adaptation of economic life to environmental constraints, this idea nonetheless plays a role in his approach, as is evident in his treatment of livestock-raising. The continuous work level associated with animal husbandry in his account differs notably from Blanchard's picture of radical seasonal shifts in workloads that can be attributed to the mountain climate.

By uniting the great variety of natural environments and forms of human existence into a single whole, ecological models often taken on a markedly synthetic character. This makes it difficult to analyze specific relationships, shifting attention away from singular assumptions and experiences. In any case, it is more important for the discussion concerning interactions between environment and development to try to reach agreement on methodological grounds than to force a consensus about the facts. In previous chapters we have examined demographic growth as a key factor of agricultural and urban development in the Alpine space. In so doing, we have tried, as far as possible, to take geographic conditions—especially elevation levels—into account. We have seen that environmental factors affect historical developments in variable ways. This conclusion sheds a different kind of light on some of the judgments and discourses that one finds in the scholarly literature about nature's impact on human experience in the Alps.

The problem with adaptation to the environment is that over time, 'adaptation' is itself adaptive in a way—it can mean different things in different contexts. A small population would use the mountain region and its various characteristics differently than would a larger population, which might produce for a long-range market. One can take as a given that with each stage of agrarian intensification preference was shown for land that appeared to be particularly adapted to reigning conditions. For example, if plant cultivation was not very widespread, then available flatter terrain was attractive. As the total size of the grain-growing and grape-growing surface increased, in some cases it was useful to build terraces on steeper slopes. When the demographic pressure continued, one could then find the intensive cultivation of bottomlands in valleys and along rivers—places that had previously been used for pasture because of the amount of labor that would have been required to improve them.

Each phase made it necessary for people involved in agriculture to evaluate the advantages and disadvantages of specific ways of using resources, but at no point is it possible to derive the mode of exploitation in any direct, fixed way from environmental and terrain conditions. In this sense, the discourse about environmental

Environment and Development

adaptation is misleading. This historically sensitive approach applies to the problems of livestock-raising and Alpiculture as well. These are often represented as paradigms of how economic activity adapted itself to mountain regions. This picture obscures the fact that livestock-raising—in contrast to plant cultivation—was carried out over the entire course of the year. Animals had to be wintered in ways that do not fit well with the notion of a "natural" economic activity.[28]

The modest resources and tight growth limits in the Alpine space require careful consideration too. Any evaluation of environmental potential is strongly colored by how the one defines the object of study. Much existing research focuses on the demography of small localities over brief periods of time, since available sources invite richly layered treatments of such phenomena. It is not unusual for these kinds of studies then to mistake the methodological boundaries of their research areas for real, effective territorial boundaries, and to measure the carrying capacity of these areas by the short-term state of population and agriculture. Thus the scarcity of natural resources and its impact on demographic developments become important topics, while demographically-motivated intensification processes receive little attention. But if one begins by framing the object of study in a different way, one's conclusions can shift considerably.[29]

For the entire Alpine space from the sixteenth through the nineteenth century, the importance of environmental potential should not be underestimated. During the first half of this period, growth in the Alps did not seem to have been different from that in the plains, and even during the eighteenth and nineteenth centuries, when growth differences between the two zones began to increase, Alpine growth continued. In comparison to surrounding areas, what was missing in the Alps more than anything else was a temporal resource: shorter growing periods linked to high elevations became critical as agrarian intensification proceeded. In this regard it could be said that dependence on environmental conditions was on the upswing, but at the same time the transportation revolution overcame other kinds of environmental restraints. Nonetheless, technical innovation did not lead to the "domination of nature" envisaged by the nineteenth century, but rather to further displacements in environmental relationships as development continued.[30]

Chapter Five

Endnotes

1   Hans Jakob Leu, *Allgemeines Helvetisches, Eydgenössisches, oder Schweitzerisches Lexikon*, vol. 1 (Zurich, 1747), 136–137.

2   Josias Simler, *Vallesiae Descriptio, libri duo. De Alpibus Commentarius* (Zurich, 1574 [facsimile, Bologna, 1970]), 65–134; Friedrich Ratzel, "Die Alpen inmitten der geschichtlichen Bewegungen," *Zeitschrift des Deutschen und Österreichischen Alpenvereins* 27 (1896): 62–88; "Dossier Alpenforschung in Europa," in *Bulletin Schweizerische Akademie der Geistes- und Sozialwissenschaften*, no. 1, ed. Schweizerische Akademie der Naturwissenschaften (1994), 57–73, here 62.

3   The paragraph that follows is based on chapters 2 through 4.

4   Othmar Pickl, "Die Nutzung der Holz-Ressourcen der Alpen durch den Floss-Fernhandel," in Bergier and Guzzi 1992, 277–283; Paolo Morigia, *Historia della Nobiltà et degne qualità del Lago Maggiore* (Milan, 1603), cited in Raffaello Ceschi, "Delitti e conflitti forestali," in *L'uomo e la foresta secc. XIII–XVIII*, ed. Simonetta Cavaciocchi (Florence, 1996), 567–578, here 569.

5   See for this topic in general Joachim Radkau and Ingrid Schäfer, *Holz. Ein Naturstoff in der Technikgeschichte* (Hamburg, 1987); M. Deveze, "Flottage et transport du bois sur les fleuves européens à l'époque moderne (Conséquences pour le développement économique)," in *Trasporti e sviluppo economico, secoli XIII–XVIII*, ed. Anna Vannini Marx (Florence, 1986), 181–89.

6   Bernd Gabriel, "Die Holzbringungsanlagen am Preintaler Gscheidl - vergessene Meisterleistungen alpenländischer Ingenieurbaukunst," in *Historische Wasserwirtschaft im Alpenraum und an der Donau*, ed. Werner Konold (Stuttgart, 1994), 239–246; for earlier techniques see for example Jon Domenic Parolini, "Zur Geschichte der Waldnutzung im Gebiet des heutigen Schweizerischen Nationalparks," PhD dissertation, ETH Zurich, 1995.

7   Rosenberg 1988, esp. 97, 105–6, 117; Andrée Corvol, *L'Homme aux Bois. Histoire des relations de l'homme et de la forêt (XVIIe–XXe siècle)* (Paris, 1987), 271–410; Anton Schuler, "Die Alpenwälder: Heilige Bannwälder oder Land- und Holzreserve?" in Bergier and Guzzi 1992, 109–116; recent research has lowered its estimate of the assumed effect of deforestation on flooding, see Daniel Brändli and Christian Pfister, "Überschwemmungen im Flachland - eine Folge von Abholzungen im Gebirge? Zur Durchsetzung eines neuen Erklärungsmusters im 19. Jahrhundert," in *Bedingungen umweltverantwortlichen Handelns von Individuen*, ed. Ruth Kaufmann-Hayoz (Berne, 1997), 50–55.

8   The map is in Fernand Braudel, *Civilisation matérielle, économie et capitalisme, XVe–XVIIIe siècle* (Paris, 1979) 1: 162; on this see Othmar Pickl, "Der innereuropäische Schlachtviehhandel vom 15. bis zum 17. Jahrhundert. Routen, Umfang und Organisation," in Vannini Marx 1986 (see n. 5), 123–46, esp. 125, 125 n. 3; id., "Der Viehhandel von Ungarn nach Oberitalien vom 14. bis zum 17. Jahrhundert," in Westermann 1979, 39–71; Bairoch 1988, 49; Mathis 1977, 44.

9   Wilhelm Abel, *Geschichte der deutschen Landwirtschaft vom frühen Mittelalter bis zum 19. Jahrhundert* (Stuttgart, 1967), esp. 171–80, 315–17; Alain Dubois, "L'exportation de bétail suisse vers l'Italie du XVIe au XVIIIe siècle: esquisse d'un bilan," in Westermann 1979, 11–38; important evidence may also be found in studies on the development and spread of livestock markets, see Anne Radeff, *Du café dans le chaudron. Economie globale d'Ancien Régime (Suisse occidentale, Franche-Comté et Savoie)* (Lausanne, 1996).

10   For Abel, such drastic price increases were an unusual phenomenon during a period of demographic growth (Abel 1967 [see n. 9], 176–77), but this appears dubious when one takes note of their frequency in our study area. Abel's interpretation is tied to a cyclical Malthusian model and overlooks the economic implications of the fact that, already during the early modern period, and

independently of the demographic conjuncture, urbanization and the division of labor could have led to increases in labor productivity; for criticism, see for example Holenstein 1996, 55–57.

11   Karl Dinklage, *Geschichte der Kärntner Landwirtschaft* (Klagenfurt, 1966), 135–137; Viallet 1993, 214; Albin Marty, *Die Viehwirtschaft der Urschweiz und Luzerns, insbesondere der Welschlandhandel 1500–1798* (Zurich, 1951), 97.

12   Livestock prices, which rose in inner Switzerland faster than grain prices, helped stimulate expanded exports; see Marty 1951 (see n. 11), 61; Dubois 1979 (see n. 9), 27–30.

13   Marty 1951 (see n. 11), 48–50; Dubois 1979 (see n. 9), 26.

14   *Atlas steirisches Bauerntum* 1976, map 35; increases in cheese production and exports were significant in the plains and pre-Alpine areas of Switzerland, see Hans Brugger, *Die schweizerische Landwirtschaft in der ersten Hälfte des 19. Jahrhunderts* (Frauenfeld, 1956), esp. 93–107; id., *Die schweizerische Landwirtschaft 1850 bis 1914* (Frauenfeld, 1978), esp. 178, 189, 235; Mathieu 1992, 249.

15   Braudel 1972, 1: 51; Cavaciocchi 1994; for research on Alpine migration, see *Col bastone e la bisaccia per le strade d'Europa. Migrazioni stagionali di mestiere dall'arco alpino nei secoli XVI–XVIII* (Bellinzona, 1991); *Gewerbliche Migration im Alpenraum* (Bozen/Bolzano, 1994); *Mobilité spatiale* 1998.

16   Fontaine 1993; for a critique of the existing scholarship, see id., "Solidarités familiales et logiques migratoires en pays de montagne à l'époque moderne," *Annales ESC* (1990): 1433–1450; id., "Les réseaux de colportage des Alpes françaises entre 16e et 19e siècles," in *Col bastone* 1991 (see n. 15), 105–129; Viazzo 1989, 121–77, 294–96.

17   Jan de Vries, *European Urbanization 1500–1800* (London, 1984), 218; Dolf Kaiser, *Fast ein Volk von Zuckerbäckern? Bündner Konditoren, Cafetiers und Hoteliers in europäischen Landen bis zum Ersten Weltkrieg. Ein wirtschaftsgeschichtlicher Beitrag* (Zurich, 1985).

18   Jon Mathieu, "Migrationen im mittleren Alpenraum, 15.–19. Jahrhundert. Ein Literaturbericht," in *Bündner Monatsblatt* (1994): 347–62, here 352–53; for the link between urbanization and migration in general, see de Vries 1984 (see n. 17), 199–231; Jacques Dupâquier, "Macro-migrations en Europe (XVIe–XVIIIe siècles)," in Cavaciocchi 1994, 65–90, esp. 80–84.

19   Viazzo 1989, 261; Albera 1995, esp. chapter 18.

20   See chapter 4 above, esp. maps 4.1 and 4.2; obviously, similarities between territories provide only an initial clue—migration processes in their full complexity can not be extensively examined here. For example, for certain regions and periods, more emphasis should be placed on push-factors.

21   Favier 1993, 349–354; for the distance model in general, see Dupâquier 1994 (see n. 18), 83–84; Christian Pfister, *Bevölkerungsgeschichte und historische Demographie 1500–1800* (Munich, 1994), 44–45, 106–7; for the same reasons the "flight from the countryside" of the nineteenth century is more easily documented for some lowland areas than for the Alps.

22   Various indications in Ceschi 1994, 15–45; Braudel 1972, 44–47.

23   The strengths and weaknesses of the argument linking density to state formation can be demonstrated for instance by evidence from the larger cities. Everywhere in our study area, these were power centers, though their relation to state power varied considerably; see Tilly 1992; Ann Katherine Isaacs and Maarten Prak, "Cities, Bourgeoisies, and States," in *Power Elites and State Building*, ed. Wolfgang Reinhard (Oxford, 1996), 207–34.

24   In what follows, I focus more on the discussion of ecological arguments in Alpine history from the sixteenth through the nineteenth centuries than on historiographic aspects; for the latter see Viazzo 1989.

Chapter Five

25   "Rien d'étonnant dès lors qu'ils soient livrés à ce que nous appelons la routine, qui n'est en fait que l'attachement à des pratiques éprouvées, imaginées par de lointains ancêtres pour déjouer les embûches d'une nature marâtre" (Blanchard, "La vie humaine en montagne," *Revue de Géographie de Lyon* 27 [1952], 216–17); also ibid., 211–17; Blanchard 1938–1956.

26   Netting 1981, citation from 42, 90; later he criticized the local equilibrium model for glossing over historical disequilibria and contacts with the outside world, see id., "Reflections on an Alpine Village as Ecosystem," in *The Ecosystem Concept in Anthropology*, ed. Emilio F. Moran (Boulder, 1984), 225–35; see also the expanded discussion in Netting 1993.

27   Mitterauer 1986; id., "Ländliche Familienformen in ihrer Abhängigkeit von natürlicher Umwelt und lokaler Ökonomie," in Mitterauer 1990, 131–45, citation from 132–33; id., "Peasant and non-peasant family forms in relation to the physical environment and the local economy," *Journal of Family History* 17 (1992), 139–59; see also chapter 8 below.

28   See also chapter 3 for the indefensible idea of transhumance as some sort of historically primitive economic form.

29   Even studies that underline the openness (rather than the isolation) of Alpine society sometimes make use of the concept of carrying capacity in order to highlight the importance of migration, see for example Fontaine 1991 (see n. 16).

30   For an example of how technical innovation led to increased dependence on the environment, see chapter 3 above, page 70.

1 The location of the Alps (see p. 9).

Europe seen from a distance of 650 km; the Alpine arc, which separates Italy from northern Europe, can be identified by the cover of snow and forests (seen from a satellite).

2 "An Italian mountain," watercolor by Albrecht Dürer, 1495.

In the Val di Cembra (Trentino). The artist from Nüremberg was a pioneer of landscape painting. Ashmolean Museum, Oxford.

3 "Map of Italy" by Jacques Signot, 1515 (see pp. 87, 206).

This military map belonging to the king of France shows Italy, rich with cities and bordered by the Alps (in the left and top margins), which had to be crossed in order to reach the contested region. Bibliothèque Nationale de France, Paris.

arte Ditalie?

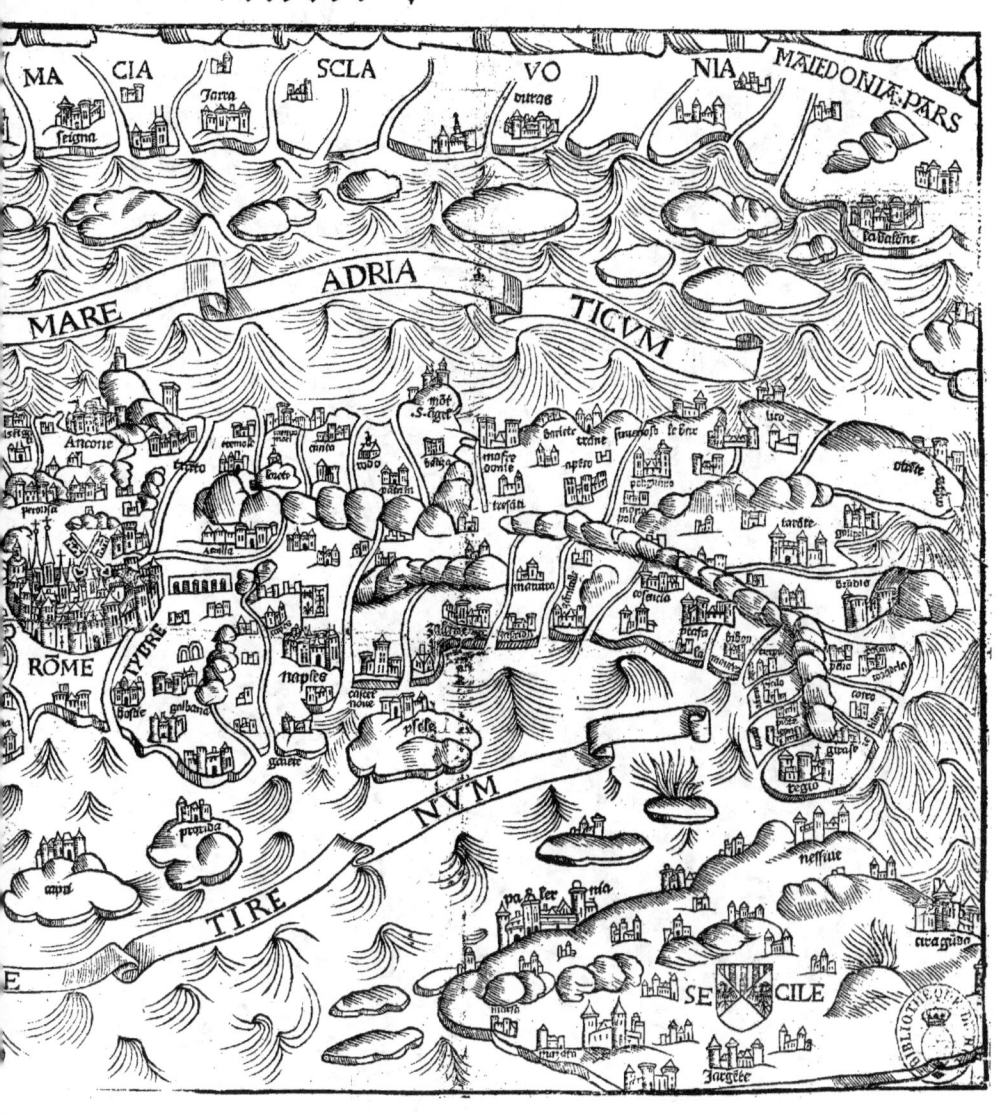

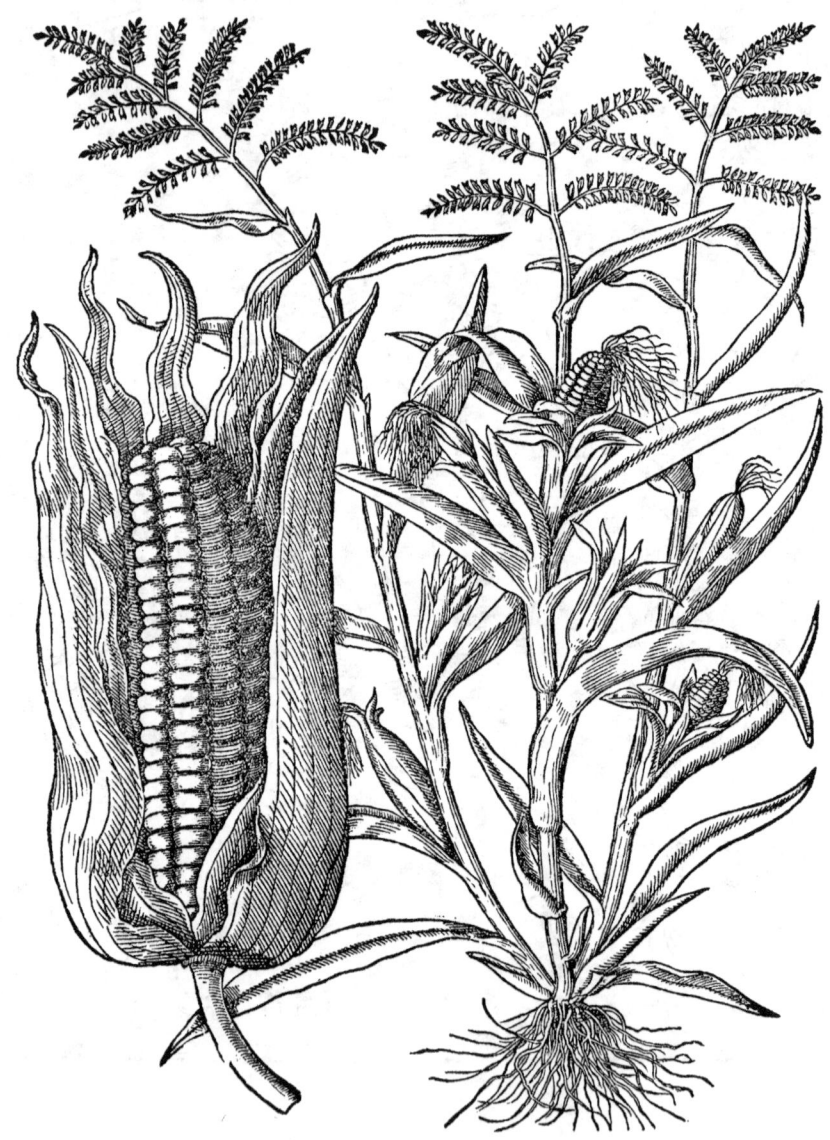

4   Maize plant, by Petrus Andreas Matthiolus, 1565 (see pp. 65–67).

Matthiolus cites the American origin of the new plant (frumentum indicum); in 1571 he also mentioned the preparation of polenta by inhabitants of the Alps. Maize cultivation, which brought high yields but was labor-intensive, became important only in the eighteenth and nineteenth centuries. Petrus Andreas Matthiolus, *Commentarii in sex libros Pedacii Dioscordis Anazarbei de Medica materia* (Venice, 1565), p. 393.

5 Homage rendered to Emperor Leopold I in Klagenfurt, 1660
(see pp. 95, 182–184).

The hereditary Hapsburg land of Carinthia paid homage to its duke. Klagenfurt, which after 1518 belonged to the Carinthian nobility and served as the site of its regional assembly, had the feel of a city of the Renaissance, even in the seventeenth century. Hans Sigmund Ottenfels, "Beschreibung oder Relation über den Einzug und Erbhuldigungs-Actum in dem Erbherzogthumb Kärndten," Universitätsbibliothek, Vienna.

6   Grenoble and its surroundings in 1604 (see pp. 89, 96).

Rapid growth in the sixteenth century made Grenoble the largest city in the Alpine space. Soon after 1600, the capital of the Dauphiné boasted new fortifications. François de Dainville, *Le Dauphiné et ses confins vu par l'ingénieur de Henri IV Jean de Beins* (Geneva, 1968), plate ix.

7 The Savoyard cadastre, 1728-38 (see p. 166).

The cadastre of the entire territory of the duchy of Savoy was, during the eighteenth century, held to be a masterpiece of geometry and state administration. In the detail illustrated here, a portion of the commune of Allèves, both parcels and mountains are represented. Archives départementales de la Haute-Savoie, Annecy.

LOI

Relative aux Droits féodaux

Du 25 août 1792, l'an 4.ᵉ de la Liberté.

N°. 2250.

L'Assemblée Nationale, confidérant que le régime féodal eſt aboli, que néanmoins il fubſiſte dans ſes effets, & que rien n'eſt plus inſtant que de faire difparoître du territoire François ces décombres de la fervitude qui couvrent & dévorent les propriétés, décrete qu'il y a urgence.

Décret définitif.

L'Aſſemblée Nationale, après avoir décrété l'urgence, décrete ce qui ſuit :

Article Premier.

Tous les effets qui peuvent avoir été produits par la maxime *nulle terre fans feigneur*, par celle de l'enclave, par les ſtatuts, coutumes & regles, ſoit générales, ſoit particulieres, qui tiennent à la féodalité, demeurent comme non avenus.

I I.

Toute propriété fonciere eſt réputée franche & libre de tous droits, tant féodaux que cenſuels, ſi ceux qui les

A

---

8   Law concerning feudal rights, France, 1792 (see p. 148).

It took the French Revolution only four years in order formally to abolish traditional lordship rights. The law of 25 August 1792, shown here as published in the department of the Isère, assigned to the seigneurs the obligation of proving their compensation claims. Bibliothèque municipale, Grenoble.

9   Hay harvesting below the castle of Hochosterwitz, Carinthia, ca. 1795 (see pp. 182–183).

In the late sixteenth century the stronghold of Hochosterwitz was transformed into a fortress resembling a palace. This image provides a dramatic contrast between the castle and the peasants at work. Lithography of Ferdinand Runk and Johann Ziegler, Vienna, by F.X. Stöckl.

10   Navigable canal near the Preintaler Gscheidl, north of Semmering, 1847
     (see p. 120).

Timber exports from the Alpine regions increased in importance after the eighteenth century.
In 1822 navigable canals were built near the Preintaler Gscheidl – they carried provisions as
far as Vienna. Illustrierte Zeitung (Leipzig, 1847), p. 72.

11 View of a hotel, Davos (Grisons), after 1885 (see pp. 105, 108).

During the late nineteenth century, health tourism experienced a favorable conjuncture. Davos, at an altitude of almost 1600 meters, quickly grew to an urban dimension. Graphische Sammlung, Zentralbibliothek, Zürich.

12 Tribute from an Alpine pasture at the church of Nendaz, Valais, ca. 1905 (see pp. 51–53).

The members of the consortium of the Alpine pasture of Nendaz deliver a tribute payment in kind (cheese) to the vicar of the commune on St. Bartholomew's Day. Alpiculture was productive, but its portion of the overall production of mountain agriculture should not be overestimated. André Guex, *Valais naguère. 281 photographies anciennes* (Lausanne, 1971), no. 99.a

13 Mountain farm at Hüttau, Salzburg, ca. 1920 (see pp. 139–145).

Family members and hired hands of the Hintersattlegg, a medium-sized farm in the Salzburg region. In contrast to the western and central portions of the Alpine arc, in the east there were many large farms with male and female hired workers. Property of M. Promegger, Hüttau.

6 Two Agrarian Structures (Nineteenth Century)

This book examines the history of the Alpine space from the end of the Middle Ages until 1900 from the perspective of two central questions: (1) How did demographic expansion, economic development, and the Alpine environment affect each other? (2) In what ways did political factors influence the structures of rural societies? Until now we have focused on the first question, which is essentially an economic one. This chapter and the next two will turn to the second problem, which involves a discussion of politics and society in Alpine history. It is undoubtedly impossible to separate issues of state development and power distribution among various social groups from their underlying economic context. Agricultural conditions were particularly important for the formation of political and social influence during our period, since the majority of the population worked in the agricultural sector, which also accounted for the majority of public spending. The *Agrarverfassung*, or agrarian structure, was in other words of central importance, situated at the nexus between the economy and politics, or between politics and the economy—the question of causal primacy must remain open for the time being.

Chapter Six

The available general literature on this theme (which is smaller than what one might expect and wish) frequently characterizes the agrarian structure of the Alps or even of mountain areas elsewhere in Europe, in homogenous terms of small farms. "Mountainous regions all over Western Europe represented remote fastnesses of small peasant property, allodial or communal, whose rocky and exiguous soil offered relatively little attraction for manorialism," writes Perry Anderson in his overview of the historical development of early modern states, with particular reference to the Swiss Confederation: "The Swiss Alps, the highest range in the continent, were naturally a foremost example of this pattern." Other authors make different claims, linking political phenomena to certain kinds of Alpine *mentalités*. For example, in the French scholarship we read: "On the whole, we can say that in the Alpine mountains we find a precocious democracy of small landowners who also hold property in common." Alongside this arrangement of small, independent, peasant landownership and shared woodlands and pastures, this interpretation attributes two apparently contradictory mental traits to Alpine inhabitants, namely: "a very strong sense of individualism and of ownership of one's own property and, on the other hand, a very strong attraction to the administration of communal affairs, economic ones, and then political ones."[1]

In my view it is both more interesting and more realistic to look at Alpine agrarian structure from the perspective of its diversity. More interesting, because this enables us to see important relationships with nearby areas surrounding the Alps, and to identify more precisely which factors had a determinant influence. More realistic, because the feudal agrarian structure of large farms, which predominated in the eastern part of the mountain system, was quite different in many ways from the communal small-farming structure described above, which often gives rise to hasty generalizations. These two forms, which we have described here only provisionally, can be taken as the main ones in our study area. But in addition to these, as will be shown, other differences can also be pointed out, such that it sometimes makes sense to speak of three or more agrarian structures, instead of two.

I am not interested in defending a rigid position with respect to these issues of definition and classification. In such a case one would have to take into account the fact that the concept of 'agrarian structure' is not at all precisely delineated, but can be construed widely to include the entire juridical and societal background of the agricultural economy. Heide Wunder, for example, "takes it to mean the whole structure of rural social power relations." A general typology would have to take numerous criteria into account, and to be attentive to both gradual transitions between forms and qualitatively different gradations. This would likely shed little light on historical processes.[2]

For our purposes, we simply need a manageable way of handling the concept of 'agrarian structure.' The farming establishment—that is, the farm itself—can serve as a starting point for this discussion, which can then follow relationships both upward and downward. Moving upward, one encounters the spheres of feudal lordship, the state, and financial organization, but also links to neighborhood and community. Moving downward, one finds questions about household organization, labor force, and inheritance. No one should think about any of these phenomena as autonomous. A *leitmotiv* of the following arguments is precisely that they are connected and constantly interacting. Existing scholarship on the Alpine space has dealt with these dimensions in an uneven manner. For some of them there have been important recent comparative studies; for others we have only a limited amount of regional or national research. One of our goals in this panoramic work is to try to balance this different scholarly production.[3]

In so doing, the historian would want to follow how things unfolded in time, and to represent history from beginning to end. In this case, however, there are good reasons for choosing a different mode of presentation, because the state of the primary sources only permits a systematic, quantitative treatment of the subject from the end of the study period onward. Therefore, we will begin by familiarizing ourselves with the situation in the nineteenth century. Then, chapter seven will describe developments over the course of the early modern period, using regional examples from several parts of the Alpine arc. Finally, in a more wide-ranging discussion, chapter eight will analyze the causes and consequences of the different sociopolitical configurations.[4]

Farming establishments

In the nineteenth century, although agrarian statistics underwent rapid development in most countries, farming establishments became subjects of investigation only at a relatively late date. Almost everywhere the analysis of property relations and of harvest status, which were key elements both for social and fiscal policy as well as for food supply and commercial interests, preceded a serious study of farm structures. In France in 1862, 1882, and 1892 data were collected on farms and how much surface area they cultivated (long after having gathered other kinds of information). In Hapsburg Austria the first real census of farms was carried out in 1902, and in Switzerland three years later. In Italy, where record-keeping in some sectors was quite up

Chapter Six

to speed with European trends, the first farm census was ordered only in 1930 (after the *Institut International d'Agriculture* had issued an appeal for such efforts).[5]

Despite this delay, it is possible to form a general overview of the situation in the Alpine space in around 1900. The Italian provinces of Bolzano and Trent belonged to the Tyrol until the end of the First World War and were thus included in the Austrian census of 1902. Comparison of these data with those collected in 1930 reveals such small changes in stratification, and the other Italian provinces displayed such unambiguous results in 1930, that in order to engage in an approximate analysis we can assign the same results to the turn of the century without risking much.

As in the preceding chapters, we underpin our study with a territorial grid that includes all of the modern, politically-consolidated administrative units of a certain size, at least three-quarters of whose surface area is located in the Alpine space. This gives us 26 regions that stretch across the entire mountain system (see chapter 2, map 2.1). For one section of the sample, though, the regional grid must be fine-tuned a bit further. In the Hapsburg Austrian area, that is, in the present-day Austrian *Länder* plus the just-mentioned provinces of Bolzano and Trent, we will base our work on the district boundaries that existed in 1902. This enables us to subdivide these rather large territories, permitting a surface area division that corresponds roughly to the size of Swiss cantons. This is also important because relations in the eastern portion of the mountains are of particular interest for the questions being addressed here. (The districts are listed in the appendix, table A.5.)[6]

Even though by around 1900 state administrations had acquired considerable experience in countrywide statistical collection projects, creating a farm census was a difficult undertaking at the time. Identifying the precise surface area cultivated by each farm—a fundamental parameter for this kind of statistical set—met with considerable obstacles since frequently not even farm operators knew how many hectares they were cultivating, whether the land was their own or rented. There were public or private documents that partly provided such information, and local units of measure could be approximately converted into hectares. But one still had to reckon with mistaken estimates for the areas of extensive exploitation that were quite widespread in mountain zones. In addition, farm operators (notwithstanding the assurances provided by data collectors) worried that the information provided would be used for fiscal purposes, and so it appears that, whenever possible, they slightly underestimated surface areas. State administrations, however, insisted on measuring the entire surface areas worked by farm establishments, including those parts which provided only small yields, such as the so-called *superficie non cultivée* (uncultivated lands) in France, including the *terrains rocheux et de montagne* (rocky and mountainous terrain) and other hard-to-define places. Woodland occasionally provided an exception to this rule. In contrast to what took place in Austria, in Switzerland autonomous forestry concerns that had no ties to agriculture were subject to a separate census. Despite all of these difficulties, the census findings still provide useful information about the situation of farming establishments. In evaluating these data, one can be reassured by the fact that an analysis carried out on the basis of widely differentiated size categories (from 5 to 10 hectares, from 10 to 20 hectares, etc.) greatly reduces the probability of distortions created by classification errors.[7]

Comparison of the findings of the censuses carried out in around 1900 does not support the picture of a unified agrarian structure stretching across the Alpine space. Although it is true that in each region one could find farming establishments of every different size, their statistical distribution varied greatly from one region to another (see table A.2 in the appendix). In the district of Salzburg, 49% of the farms had 10 or more hectares under cultivation—we will call these mid-sized to large establishments—while 37% were smaller farms of 5 or fewer hectares. In the Trentino, meanwhile, these smaller farms represented, with 91% of the total, the overwhelming majority, while mid-sized to large farms only accounted for 3%. In the most western Austrian *Land*, the Vorarlberg, the latter categories were 14% of the total, roughly the same percentage as in the neighboring cantons of eastern and central Switzerland. If we complete the Swiss census by including an estimate for farms of up to half a hectare (this smallest category having originally been uncounted), we find that 10–17% of the farms in these cantons were mid-sized to large. Of this group, a considerable number were collective establishments of sprawling Alpine pastures that were naturally registered—in the same way as establishments with individual owners—as autonomous economic units. (For this reason, in some regions the figures for the largest category, of 50 hectares and above, are higher and not lower than the figures for the next smallest category of 20 to 50 hectares). Almost everywhere else in the western Alps and on their southern slope the structure of small farms clearly predominated, with mid-sized to large farms accounting for between 3% and 8% of the total. The French departments of the Hautes-Alpes and especially the Alpes-de-Haute-Provence were an exception, with clearly higher percentages (18% and 21%). An analysis of the type of cultivation carried out in these areas shows that the size and the distribution of the *superficie non cultivée* played an important role in this regard.[8]

Chapter Six

Map 6.1: Medium and large farms in Alpine regions and districts, 1900

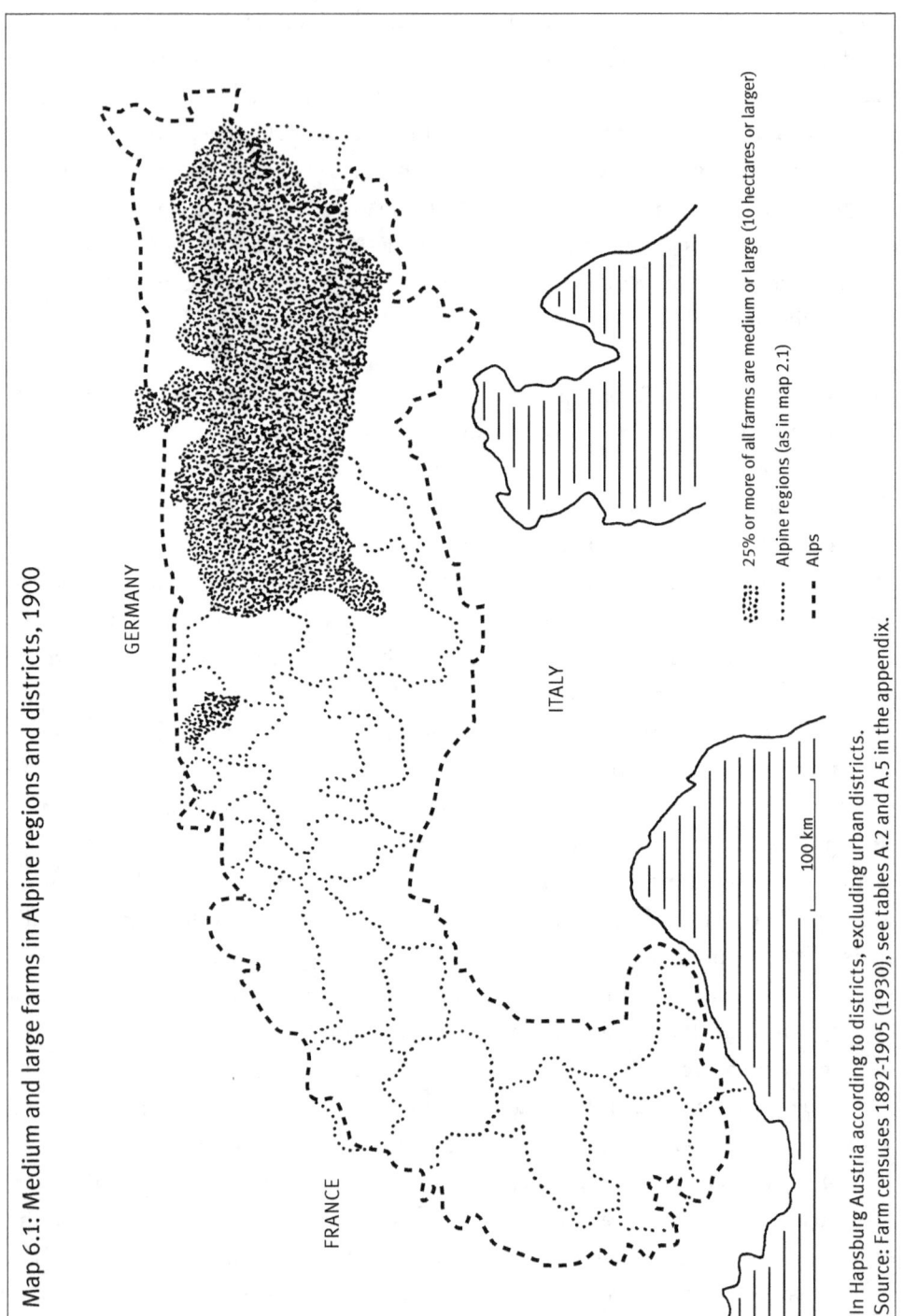

In Hapsburg Austria according to districts, excluding urban districts.
Source: Farm censuses 1892-1905 (1930), see tables A.2 and A.5 in the appendix.

Two Agrarian Structures

Map 6.2: Agricultural servants in Alpine regions and districts, 1900

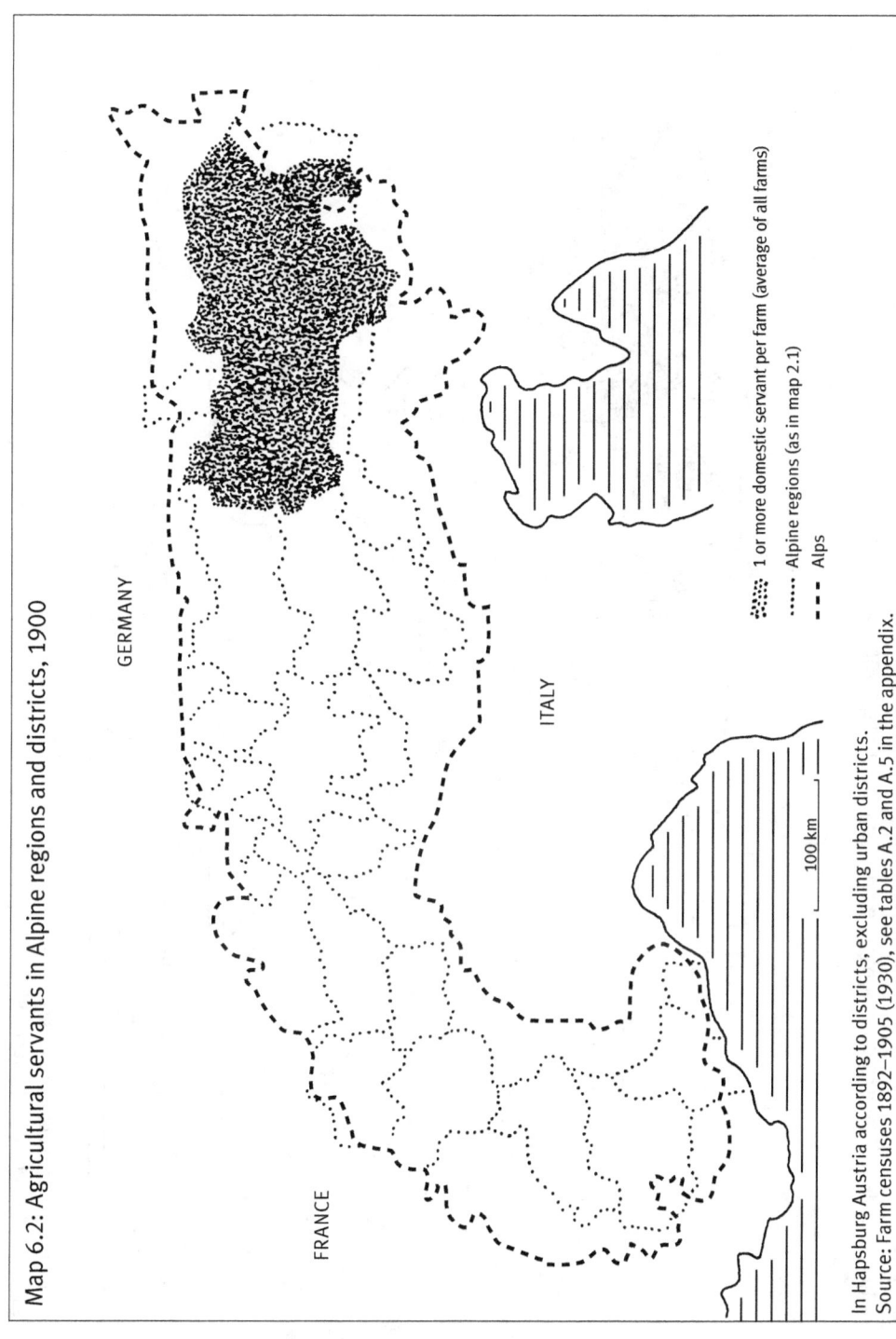

In Hapsburg Austria according to districts, excluding urban districts.
Source: Farm censuses 1892–1905 (1930), see tables A.2 and A.5 in the appendix.

Chapter Six

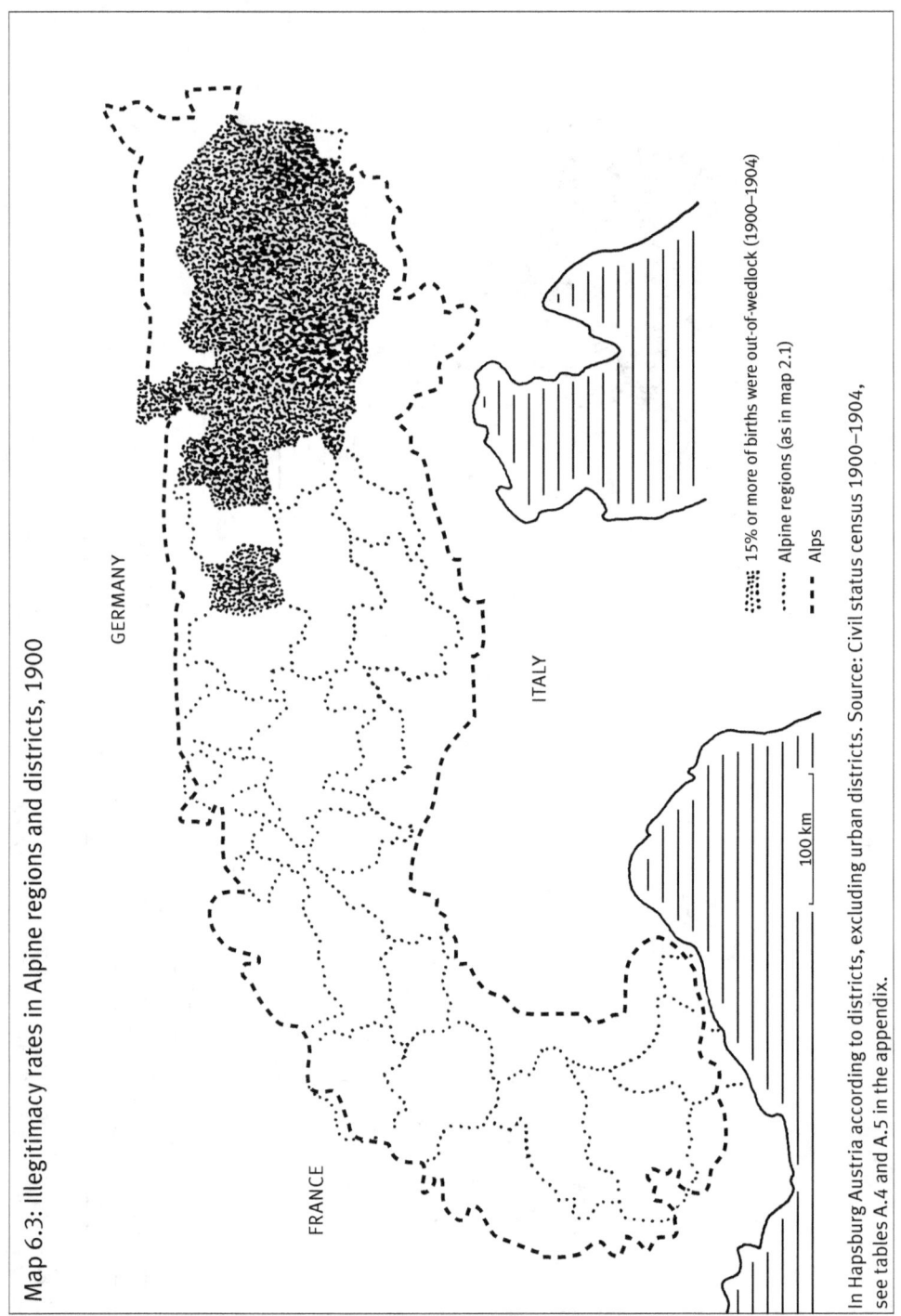

Map 6.3: Illegitimacy rates in Alpine regions and districts, 1900

In Hapsburg Austria according to districts, excluding urban districts. Source: Civil status census 1900–1904, see tables A.4 and A.5 in the appendix.

Still, these French enclaves did not reach the size of many Austrian Alpine regions. If one isolates the percentage of mid-sized to large farms as the basis for measuring stratification, fixing the figure of 25% as a dividing line, one obtains the image represented in map 6.1. The entire northeastern portion of the study area, from Styria halfway across the Tyrol, along with the small detached district of Bregenz in the Vorarlberg, bordering on Bavaria, appears here as a zone with a high percentage of larger farming establishments. The area is almost totally compact and without spatial interruption, even though we took care to subdivide this area into districts rather than using the larger present-day Austrian *Länder*. In order to indicate how broad the differences are, it might be useful to mention some averages. In the *Land* of Salzburg in 1902 the average size of a farm's cultivated surface area was at least three times larger than in the Vorarlberg—and at that time, as we have mentioned, the Vorarlberg was by no means the Alpine region with the smallest average farm size. If we extend this comparison beyond the borders of Hapsburg Austria, although this admittedly increases the statistical uncertainty, we find that the average farm in one region could be up to ten times as large as in another region.[9]

Similar differences become apparent when one examines the agricultural labor force on the basis of the same data. Of particular interest are the non-family members working in the farming establishments (see table A.3 in the appendix). To be sure, owing to the sporadic nature of their work, it was difficult to identify consistently the men and women who were employed as day laborers. Incomplete and uncertain statistics can only indicate that in some regions on the southern slope of the Alps day labor was more widespread (Alpes-Maritimes, Imperia, Belluno). However, regularly employed farmhands, both male and female, were integrated into the household and could be counted more easily. The average number of permanent hired hands per establishment was 1.2 in Salzburg and 1.3 in Carinthia, but only 0.1 in the Trentino and 0.2 in the Vorarlberg. In some Swiss cantons the figure was slightly higher (0.4), but here the overall picture was influenced by two facts: the smallest category of farms, those with the least amount of hired hands, was not included in the census; and the large pasturing consortia hired numerous workers for their Alpine pastures. Otherwise, everywhere else the numbers of hired hands were minimal, even in the two French departments with relatively large average farm sizes.

Map 6.2 shows those areas where farms employed, on average, one or more hired hands. This gives us a picture of a zone in which hired agricultural workers were widely diffused, from Styria to the eastern Tyrol. If one were to reduce the limit to an average of somewhat less than one hired hand per farm, a whole series of other Austrian and South Tyrolean districts would be added, rendering the map more precisely congruent to the previous one. It might be that the baseline number of hired workers mentioned here does not appear very large, but one must keep in mind that if the smaller establishments were to be excluded from consideration, one would be left with much larger average numbers of hired hands. In Carinthia, for example, the average for farms of 20 to 50 hectares was 2.3, and for farms of 50 hectares or more it reached 5.2. On certain individual farms, the number of hands was 10 or even higher. What was characteristic of these eastern Alpine regions was the fact that large farms such as these, with numerous hired workers, figured prominently within the entire scheme of agrarian stratification.[10]

Another indication of the peculiarity of the eastern Alpine area is provided by the illegitimacy rates that can be calculated on the basis of demographic statistics. In contrast to the indicators that we have been examining heretofore, these rates are not limited to the agrarian sector, but refer to the entire population (see table A.4 in the appendix). In the five years from 1900 to 1904, the percentage of extramatrimonial births recorded in Salzburg was 26%, in Styria it was 31%, and in Carinthia it was even 40%. In the Trentino, on the other hand, it was only 1% and in the Vorarlberg it was 6%. For most of the other regions in our sample the percentages were between the last two figures mentioned. Higher percentages were to be found only in the Val d'Aosta and in the Alpes-Maritimes (10% and 12%). If we identify the regions and districts where 15 or more out of every 100 births were illegitimate, we obtain map 6.3, which once again coincides to a large degree with the maps of the preceding indicators. The relationship between married and unmarried births was of course determined by a wide variety of factors. In the Maritime Alps, where the illegitimacy rate doubled in the three decades prior to 1900, increased urbanization and mobility along the Mediterranean coast seems to have played an important role.[11] The rates in Austria were higher by far, very high even in pan-European comparison, and were rooted in the Alpine agrarian structure of large farms. The strata of the rural population that was excluded from positions of farm management—hired hands, but also workers who were subordinate family members—produced the majority of out-of-wedlock births.[12]

On the whole there are good reasons for distinguishing the structure of eastern farming establishments from that of western ones. In this context it is also important to observe that in the eastern regions a relatively high number of properties were isolated farm complexes, a fact that was already known through systematic data collection before the turn of the century. These isolated farms also included pasture and wooded areas that often formed part of the village commons when such agglomerated settlements were present.[13]

If we do not take into account the subregional district boundaries and make a completely schematic comparison between the regions of large farmers—Styria, Carinthia, Salzburg, Tyrol, and Bolzano—and the other regions in our sample, we find the following characteristics: in the 'east' the percentage of mid-sized to large farms was four and a half times larger than in the 'west'; the number of hired workers per farm was at least seven times as high; and the illegitimacy rate was at least six times greater.[14] If we look at the district data, though, we find a notable level of short-range variation, and also that our indicators do not always correlate rigidly to each other. For example, in St. Veit and in Murau, two contiguous districts in Carinthia and Styria, the indicators were high, without exception (53% and 62% of mid-sized to large farms; 2.2 and 2.5 hired workers per farm; and 48% and 57% illegitimacy rates). On the other hand, in some districts the indicators varied, especially in the Tyrol and in Bolzano (the South Tyrol), which the data reveal to have been truly transitional regions between east and west. Short-range variation also reflected long-range variation here (see table A.5 in the appendix). If one takes as a point of reference the district subdivisions and the various indicators along with the threshold limits established above, a measurement of the surface area of the 'eastern-like' zone shows it to be roughly a quarter to a third of the entire sample's area. From the perspective of the Austrian historiography, this area is large enough for its agrarian structure to appear to be the norm for the Alps. From the point of view of research across the entire Alpine arc, though, it must be considered an important special case.[15]

The sources that provide information on the situation prior to 1900 generally have more holes and are for the most part comparable only under certain conditions. For example, references to occupation in any of the national population censuses carried out from about 1870 onward help show how the main centers of demand for day workers, especially in the agrarian labor sector, were found in the same regions described above. But because of the problematic and sometimes varied collection methods used in the censuses, it is difficult to calculate trends on the basis of these data.[16] Only the indicator for illegitimacy rates, which had long been standardized because of its normative use for state and church, could be employed for quantitative comparisons across the entire study area during this period. The changes that took place between 1870 and 1900 were often limited and contradictory: in half of the regions the percentages increased, and in the other half they declined. Still, the shifts were not distributed uniformly across the entire sample. In the regions with modest illegitimacy rates, the tendency was toward slight increases, while the regions with high rates showed, on the whole, clear signs of a decrease, such that for the late nineteenth century one can speak of a leveling-out between east and west (see tables A.4 and A.5 in the appendix).

At the beginning of the nineteenth century the tendency in the east had been heading in a different direction, though. Studies from the Austrian side have shown that illegitimacy rates increased, sometimes rapidly, until about 1870. These increases are mainly tied to the agrarian revolution of the period; that is, to intensification processes prior to the beginning of markedly industrial agricultural mechanization. During this phase, farms needed extra labor, a fact that led not only to the increased employment of agricultural workers, but also to greater reliance on the labor of children and relatives. In this way, the social groups that were especially important for the increase in out-of-wedlock births expanded (ahead of the point when, late in the century, a decrease in the number of hired hands, along with other factors, reversed this tendency). The need for agricultural labor seems to have been so great in some areas that the procreation of illegitimate children was tolerated by large farmers and sometimes even promoted.

Nonetheless there are also indications that illegitimacy in the Austrian territories had already reached high levels in the centuries prior to the 1800s. This requires us to shift our search for explanatory factors away from medium- and short-term phenomena toward regional continuities. From a competent source we learn that "the structural conditions surrounding the high illegitimacy rates in the central part of the eastern Alpine space reach far back in time," and that this applies for conditions such as settlement structure, the relatively large farm complexes, and the situation of hired agricultural workers.[17] Reverse indicators show that this was also true, as we will see below, for other regions in the Alpine space.

Public order and property

Beginning in the eighteenth century and with the French Revolution in particular, the impact of wide-ranging state forces on the organization of lordly domination and on rural landowning increased rapidly. In order to understand the shift in this important aspect of agrarian structure, mountain regions should not be abstracted from the political contexts that included them, or into which they were added.

Two Agrarian Structures

Short- or long-range factors could still be caught up together. For example, the 590 inhabitants of the commune of Avançon, near Gap in what is today the department of the Hautes-Alpes, experienced the start of the French Revolution in the second half of April 1789. Food supplies were low and prices high, so the subjects of the local lord decided to take back the payments in kind that had been collected from them during the previous year. They were expecting the Estates-General, which was soon to meet at Versailles, to free them definitively from these kinds of feudal dues. On April 20 the villagers, together with other people from the valley, formed an armed group and gained entrance into the nearby castle of the *seigneur*. The servants of the absent lord were forced to promise in writing that the lord would produce, within days, a written renunciation of all his rights. For his part, the nobleman had no intention of granting this request, but despite the military action that followed, he was unable to collect from his peasants the new payments that were past due. As Georges Lefebvre points out, this conflict in a circumscribed area within the Avance valley resembled a classic peasant revolt in many ways.[18] But the fact remains that in the spring of 1789 these events were accompanied by extraordinary circumstances that began to accumulate and that eventually turned the whole country upside down.

In conjunction with planned tax reforms, the king had called the first meeting of the Estates-General since the early seventeenth century. He had also asked for a study of his subjects' valid complaints. The election of the deputies for the Third Estate and the compilation of the *Cahiers des doléances* took place with the surprising broad, quasi-democratic participation of the population. In the local assemblies there was already some anti-noble unrest. In Provence, a series of riots followed by real uprisings took place in March—these probably influenced the events in Avançon. When the situation in Versailles and Paris became more acute, the provinces again reacted in a way that demonstrated the political cohesion of the kingdom. After the assault on the Bastille on July 14, local fears of brigands and thieves, and of hostile nobles and foreign invaders, merged together in the "great fear" that spread like wildfire across vast regions of the country. When rumors of some danger arrived, bells were rung furiously, women and children were hurried to safety, men formed up into companies of militia: in essence, a state of local siege was proclaimed. In the Dauphiné it appears as though a false message that arrived from the north, or perhaps an insignificant incident along the border with Savoy on July 27, was the spark for the panic that broke out. The fear reached Grenoble on that very evening, arrived at Valence on the next day, at Gap on the day after, and then at Briançon two days after that. With the exception of the Oisans, more or less the entire Alpine region was struck. This wave of fright also spread further south, affecting Provence in like manner, until August 4.[19]

In the night of that same day, pressured by events and taking advantage of the opportunities that presented themselves, the National Assembly in the capital declared that "feudalism" in all of its forms was abolished—a gesture of enormous symbolic and political relevance (discussions about "feudalism" had been ongoing for decades). When the national state of alarm calmed somewhat, especially among those who counted themselves part of the Third Estate, this principle was transposed into more restrictive terms. The decrees issued in following months abolished across the board all "personal" feudal rights (linked more to the status of one's person), while "real" feudal rights (linked more to the status of the property in question) were held to be, in the absence of contradictory evidence, contractual property titles for which recompense would be claimed. Because of the variety of such obligations and of rural relationships, conflicts over interpretation were a foregone conclusion. The principle of compensation remained controversial, as many sectors of the rural population took land rents and other land-based dues to be feudal usurpations. These debates were not resolved juridically, but politically: the course of the Revolution in the cities often paralleled that of the countryside. After the assault on the Tuileries and the proclamation of the Republic, a series of decrees were issued in fall 1792 that were basically advantageous for rural groups, above all because they required old feudal lords seeking compensation to bear the burden of proving their claims. And in July 17, 1793, both internal and external pressure brought an end to pending cases. All feudal rights that had not been compensated were abolished by decree.[20]

Agrarian reform was a central issue in the modernization of state and society, and during the period that followed it was confronted by every country in the Alpine arc. In France this reform played out very rapidly—only four years separated the Declaration of 1789 from the final abolition of feudal rights—and in terms that were very favorable to peasants. As far as the key points are concerned, almost no other place could match this development. Still, comparing different kinds of provisions produced in specific situations is difficult, and one should not assume that the French Revolution invented such reforms. From a very early stage in the northern parts of the Alps, systematic agrarian reforms were carried out in the lands ruled by the house of Savoy. After many earlier attempts, the royal court in Turin issued a decree in 1771 that emancipated inhabitants of the province of the same name from feudal obligations. This became famous everywhere, and was held up as an example precisely by reform-minded circles in neighboring France. But the abolition of feudal rights in the Savoyard lands, which proceeded slowly because of the cost involved, was completed only when French Revolutionary armies entered the territory and organized the National Savoyard Assembly in 1792.[21]

Demographic and economic expansion provides one reason for the general acceleration of the reform process beginning in the late eighteenth century. Agricultural intensification and increasing market integration required a redefinition of the bewildering variety of claims made on land and its output. The ideal for many reformers and landowners was private property that was freed from any limits on its transfer and clearly separated from the public or lordly realm. In reality, though, public order was and remained an important factor in the evolving definition of property. This was already apparent in the fact that agrarian reforms were often carried out as a result of revolutionary pressure—in 1789, 1830, and 1848. Long-term developments did not tend in the direction of absolute individual ownership, but toward an interweaving and displacement of rights over land. Local restrictions were followed by macro-level ones, which in some cases grew to the point of placing authority in the hands of the nation-state.[22]

In early 1792 it was clear that armed conflict in Europe was practically inevitable, but no one could have guessed what terrifying dimensions it would assume. In April France declared war on the emperor in Vienna, and in the fall republican troops marched into Savoy, Nice, and other territories. The Italian campaign of 1796–97 gave birth to the Cisalpine Republic in Lombardy and the Ligurian Republic in Genoa. In 1798 French armies first occupied the Swiss Confederation, and then Piedmont. Additionally, within our study area, the following territories fell under direct or indirect French control by 1810–12: the Veneto, Illyria (with parts of present-day Slovenia, Carinthia, and Upper Austria), and Bavaria (which now possessed the Tyrol and Salzburg). Following this expansion, almost the entire Alpine space either became part of Napoleon's *Grande Nation* or was dependent on it. The only exceptions were the Alpine territories of Lower Austria and of Styria. Even though the Empire collapsed soon thereafter and many boundaries were re-established in their prior locations, the French period still represented a break everywhere.[23]

What effects did these events have on the process of agrarian reform? It is impossible to provide a single response to this question. In some territories, in certain respects at least, there was clear acceleration, such as in the case of Savoy mentioned above, or in the juridical systematization produced by the spread of the *Code civil*. With

Chapter Six

militarization and expansion, though, France rapidly acquired conservative characteristics and lost interest in creating breakthrough legislation. The revolution had deeply shocked the European nobility, leading it in some cases to block proposed reforms for a long time.[24] The variety of presuppositions, influences, and reactions related to these developments makes it necessary for us to look at one area at a time.

In the Swiss Confederation, social unrest and the traditionally strong influence of France led in many areas to the elimination of old governing authorities even before the foreign troops arrived. The Helvetic Republic, established in spring 1798, introduced an entirely new conception of the state as centralizing and egalitarian. It also swiftly took up the agrarian question. Both the constitution and legislation provided for the free abolition of all "personal feudal rights" (which were practically insignificant since they were not widespread and of barely any economic consequence, but which were still thought to be humiliating for subjects and symbolically important). In addition, the rule of being able to redeem any other kind of property-related obligation was put into effect, and most beneficiaries of these dues had to be compensated. The Helvetic Republic only lasted for a few turbulent years. With the conservative restoration, control over agrarian policy was reacquired by the cantons, but the modern principle of being able to redeem dues within an unlimited time period remained intact. Following the political conflicts of 1830 and 1848, many cantons lowered prices and made it obligatory to redeem these dues. In this regard the agricultural conjuncture became more important, since the sums required for these redemptions were considerable and peasants' debt burdens increased as a result. For many peasants this created new ties of personal dependence, while the cantons were able to spread their financial structure across a wider foundation, something that was probably positive for the over-mortgaged agricultural sector as a whole.[25]

In the mountainous areas of Switzerland, the redemption issue was less important than in the plateau region. Swiss administrative records, which though incomplete were not without method, show that the aggregate of property-related obligations in the two areas in around 1800 was clearly differentiated. If calculated on the basis of the population, documented revenues from tithes and land rents in the Alpine cantons amounted to less than 70 *Rappen* per person, while the average for the cantons of the *Mitteland* plateau ranged in most cases between 250 and 350 *Rappen*. Still, such payments varied greatly from one locality to another and their low average amount does not mean that the redemption process proceeded quickly. In the Valais, for example, redemption seems to have taken place in fits and starts, above all because the church, the principal beneficiary of these obligations that were scattered across the territory, opposed any change in their form. There, some kinds of "perpetual" dues remained in place even after the First World War.[26]

In Hapsburg Austria the official freeing of rural property from these kinds of obligations (*Grundentlastung*) took place in the years following 1848 and had an impact, even in the Alpine areas, that was incomparably greater than in the Swiss Alps, because feudal rights on land and persons were generally much more prominent and extended to entire regions. In order to increase the direct influence of the state over peasant subjects, the imperial government had issued a series of laws beginning in the eighteenth century, culminating in the Josephine edict on arable land, land registers, and related revenues and dues, dated 10 February 1789. This law would have led to a reduction of the tributes owed to landowners, and to the monetization of contributions in kind and of labor dues. This provision met with such vehement resistance from noble groups that it had to be withdrawn. When an imperial decree was issued nine years later permitting the voluntary redemption of such dues there was little reason to object, since these deals were negotiated according to local interests anyway. Not least because of the French Revolution, the Austrian crown now assumed a cautious position on the agrarian issue for about fifty years. But the position and disposition of the nobility began to change. In 1815, in territories previously controlled by the French, noble-dominated provincial diets were restored, while jurisdictional lordships remained in the hands of the state. In a context of economic and demographic expansion, the great landowners became more attuned to the market and started to question the economic utility of unpaid labor, the so-called *Robot*, or corvée labor. In the 1840s some states in the empire saw the emergence of an elite opposition movement that made its case with similar modern arguments. Before rapid change could occur, though, a bloody eastern regional revolt (1846) and a pan-European revolution were necessary. Soon after March 1848, when it appeared for a time that the great Hapsburg monarchy was about to dissolve, the time had come.[27]

According to many observers, the attitude of rural groups had a decisive impact on the attempt of the educated, well-off bourgeoisie to carry out this revolution. Agrarian reform was among the key issues addressed by the representative assembly that met in Vienna, which passed a wide-ranging law on redemption that was promulgated by edict on 7 September 1848. When Imperial troops took the capital in late October and

Chapter Six

the revolutionary assembly was first recessed and then dissolved, the reform law was not abandoned. The monarchy, in the person of the newly-enthroned, eighteen-year-old Franz Joseph, integrated the law into other legislative provisions in the course of the following year, ordering its rapid execution.

"Subject status and relations of lordly protection, together with all of the laws that regulated these relations, are abolished," read the first article of the September decree. Only the second article, which dealt with the redemption of rural possessions from those who claimed landed feudal rights, was explicitly economic. The execution of the law showed how closely linked property relations were to public order. Revenues stemming from personal subject relations and from police and judicial authority were abolished without indemnity. For most other dues a "modest ransom" was to be paid. Part of the estimated value of these dues was paid by the state, part by those who were being liberated, and part of the cost was carried by the ex-beneficiaries who were giving up their former revenues. With the help of a network of commissioners and state officials the main phase of this wide-ranging redemption law was carried out within a few years, and the last obligations were redeemed before the end of the century.[28]

The territories of the Austrian Hapsburg crown that were wholly or in part Alpine were affected in different ways by the elimination of feudal dues on rural property. Labor dues and their abolition were important only in the eastern belt—Carniola, Carinthia, Styria, and Lower Austria. In these regions, documented forced labor obligations were minimal in Carinthia (five days of labor per year on average), the only completely Alpine region within this group. The opposite was true in terms of the total owed per tributary, which reached its highest point in Carinthia (with average annual payments of 9 florins). There, payments in kind and in cash were both extraordinarily high. An unusual position within the Austrian Alps was assumed by the sprawling territory of the Tyrol with its various parts (northern, southern = Südtirol/Alto Adige, and Italian = Trentino) and by the Vorarlberg. According to Otto Stolz this peculiarity resided in the fact that "in the Tyrol, landed lordship had long been experienced not in terms of subject relations, but merely as an expression of private law." Thus, for many the elimination of feudal dues was not an urgent priority. The difference, however, when compared to eastern regions, had to do less with average total burdens than with the distribution of those burdens. In the Tyrol, a much higher percentage of the inhabitants owed dues (30–57%, compared to 11% in Carinthia). In other words, a large number of small or very small dues were owed on a large number of small units. These burdens were distributed irregularly across the Tyrol. In per capita terms, redemption costs were three times less in the northern Tyrol and the Trentino than in the Südtirol/Alto Adige. The costs in the Vorarlberg, the region bordering on Switzerland, were even three times less than that and therefore the lowest by far.[29]

The main significance of all of the aforementioned agrarian reforms, which had their start in the late eighteenth century, is sociopolitical in nature. They were characterized by a formal separation of status-related relations from more strictly economic ones, helping to procure for the state a legally undifferentiated body of citizens. The enlargement and intensification of the political sphere was a prerequisite for this process. Prior to this period, in some more limited contexts, relations of personal dependence had been brought to an end through the payment of indemnities on behalf of serfs to lords.[30] Under the wide-ranging egalitarian pressure that accompanied the revolutionary waves, wherever these 'personal' relations could be found they were separated from their economic elements and abolished without compensation. The price of redeeming "real" feudal obligations, however, varied depending on the intensity of lordly domination and of state policy. In this regard there were huge differences between the French, Swiss, and Austrian regions described above.[31] These reforms did not necessarily result in economic relief for the peasants. When indemnities were paid to lords, peasants' expenses increased in the short term, and over the long term it was possible for their tax burden to grow more than their expenses diminished. The most important economic effect of these reforms lay in the creation of a more active market thanks to the elimination of unlimited dues, and to a corresponding monetization.

Particularist elements could still be identified in late nineteenth-century sharecropping arrangements, in which a certain percentage of the harvest—often a half—was owed to the landowner. Many reformers believed that sharecroppers had no incentive to innovate, and they saw this proportionality and the fact that the obligation was often paid in kind as contrary to their notion of a modern economy. Because they were valid for fixed terms, though, these contracts could not be annulled for the time being without explicitly undermining property rights altogether. In around 1900 sharecropping and related land tenure arrangements were present everywhere, but to a particularly significant degree only in southern areas, such as the Bellunese, the Trentino, and the Alpes-Maritimes. Much more important, because they were more widespread, were term rents with fixed payments, another arrangement that had long been used. But on the whole, in few regions did these various kinds of rent structures account for more than 10–15% of the cultivated territory.[32]

Chapter Six

Inheritance law, collective resources

The unification of rural inheritance law, another key element of agrarian structure, was strongly stimulated by the *Code civil des Français* of 1804, drafted under the direct control of Napoleon and then diffused widely as both a positive and negative model. The *Code civil* in fact represented a compromise between a variety of regional and social juridical conceptions from across France. It required that property be inherited "without regard to gender or to primogeniture," and "in equal parts," according to one of its over 2,000 articles. In other articles, this principle of equality was counter-balanced through an underscoring of paternal authority, considerable freedom in the writing of wills, and many open-ended formulations. With respect to the division of land, the *Code civil* recommended against the fragmentation of farms, as far as possible.[33] The actions of revolutionary French legislatures and the subsequent *Code* of the First Consul (later renamed after Napoleon himself) formally integrated the French Alpine areas into a national framework of inheritance law, while in other countries a similar process was carried out over the course of several generations.

In the Grisons local inheritance guidelines remained valid until after the Napoleonic period, when the first steps toward codification were made. In 1831 a jurist tried to organize the accumulated mass of articles issued by the Leagues and by communes. He reached the conclusion that in this barely cohesive state, there were eighteen different sets of inheritance law (which nonetheless agreed on the basic norms of equal partible division among all children and of strict limitations on the testator's discretion). Only in 1862 did a civil code take effect in the Grisons, and another fifty years passed before this code became part of Swiss federal legislation. This national legislation differed from its local and cantonal precursors mainly in the special concern shown by its provisions to the agricultural economy, permitting one of the heirs, usually a son, to exert a right of undivided inheritance of the entire farm, subject to a cash compensation of the co-heirs.[34]

In some Austrian regions the expression *Erbrecht* (inheritance law/right) referred, in early periods, less to the form of inheritance than to its existence, i.e. the right to inherit a farm. In order to obtain this right, one had, in certain circumstances, to pay an acquisition fee to the lord whose jurisdiction extended over the land in question. Thus this right also became known as *Kaufrecht*, that is, "right of purchase." As early as the eighteenth century the imperial government had not only promulgated laws limiting such feudal authority, thus favoring the regularization of inheritance law, it also issued regulations concerning the indivisibility of farms, whether among living kin or via inheritance. The general code introduced in 1812, the *Allgemeines bürgerliches Gesetzbuch*, which was an Austrian version of the codification of civil law, explicitly preserved these provisions for agricultural property. They also remained valid through the course of the standardization of property status as rural feudal obligations were being redeemed. It was only the liberal legislative activity of 1868 that abolished restrictions on property divisions in most of the crown's Alpine territories and gave owners the right to decide—although this development immediately spawned a counter-movement aimed at establishing new kinds of specific regulations.[35]

During this period, similar movements embracing ideologically conservative forms of modernization and celebrating the *Bauernstand* (peasant estate) became influential in many countries. They prompted intense debates over rural inheritance law, leading to broad investigations of rural inheritance practices beginning in the late nineteenth century, and making it clear that in some regions—though not all—there was a considerable gap between laws and actual practice. The results of these inquiries show that within the Alpine space the entire spectrum of inheritance practices could be observed, from rigidly egalitarian partition of all goods to the prominent priority of one heir. The territorial breakdown of these variations was less irregular. In the east, from the eastern Tyrol to Carinthia and Styria, the undivided inheritance of farms predominated. In almost all of the other regions one found forms of the partible inheritance of property, defined and practiced in various ways.[36]

After examining the lordly and familial dimensions of landowning structures, a third and final aspect must be mentioned: the collective rights of the inhabitants of a locality to use and own property, especially pastures and woodlands. In the political and scientific discussions of the nineteenth century, the Alpine space served as a certain reference point for those interested in alternatives to private property. Growing theoretical criticism of this institution, especially in the second half of the century, increased interest in the actual experiences of those who held property in common. Unorthodox intellectuals frequently cited examples from mountain regions, representing them as proof of a collective property regime that had been widely diffused in earlier times and that, under certain conditions, provided political direction for the present and the future.[37]

It is true that communal forms of ownership and land use in the Alpine space were important in particular ways during the nineteenth century. But rather than viewing these phenomena as evidence of a timeless tradition, or as constituting a generalizable

Chapter Six

model of social improvement, it makes more sense to see them in economic terms, as the outgrowth of an Alpine agricultural structure that was, on average, extensive. When rapid intensification was carried out in river bottoms and in areas near settlements, communal lands were frequently divided. In addition, the overblown conflict between collective property and private property did not correspond to the variety of regulations that were actually in force. Intermediate forms of property, partly familial and partly based on neighborhood, were widespread—in the small settlements of the Italian Alpine valleys, for example. In the nineteenth century, with the consolidation of communal organization, these forms fell under the growing influence of the communes, a development that often carried years of conflict in its wake. The elaboration and subsequent evolution of public property thus also depended on political dynamics. This is abundantly apparent as far as forest ownership is concerned. Despite significant opposition that sometimes went as far as civil disobedience on the part of mayors and even peasant revolts, forest lands in France underwent a swift process of nationalization after the 1830s and 1840s. The state also played an important role in the forest sector, along with landed lords, in eastern Austria. In the second half of the century, many of these lords were able to free themselves, relatively inexpensively, from peasant claims of forest rights. Thus they became, in the face of popular protests, modern owners of large estates.[38]

Looking back at this chapter, we can conclude that it makes little sense, in light of existing scholarship, to speak of a uniform Alpine agrarian structure, even if toward the end of the study period a certain alignment can be identified. Beginning in the eighteenth century, agrarian reforms on a vast scale resulted in the modernization of the property structure throughout the Alps. Its chief significance lay in the formal separation of more status-related aspects from other more economic aspects of feudal authority, a separation designed to obtain a single set of state citizenship rights. These new rights enabled the rural population to be seen in an entirely different light: the "peasant order" was transformed into the "heart of the nation," expanding its prestige in ways that definitely influenced the image of the Alps, a relatively rural region.[39] But the chronology and the import of agrarian reform differed notably in the various parts of the mountain system. In Savoy, the general emancipation decree was issued in 1771, while in Hapsburg Austria the law abolishing serfdom and lifting obligations on feudal possessions was only approved in 1848. In this eastern area, the creation of a unitary state brought about the greatest change, since the landed lords had previously exercised the greatest amount of public influence.

With respect to farming establishments, which were only partially affected by such changes, the differences between west and east were located in the respective percentages of small and large farms. Statistics show that small farms were the norm in the western regions of the Alpine arc in around 1900, while the eastern regions counted a significant percentage of large farms with many hired workers. The area dominated by large farms such as these overlapped with the area dominated by landed lordships, extending from the central Tyrol eastward, and comprising from a quarter to a third of the entire Alpine surface area. The percentage of the labor force that was excluded from farm management was much higher here than in the west, and this was reflected in the high illegitimacy rates as well. If the authority of the head of the household is taken to be a distinctively feudal characteristic, then the feudal order remained widely in effect at the end of our study period—especially in the east, where there were not only many hired agricultural workers, but where even workers who were family members were for a long time officially described as 'servants.'[40]

Generally, it seems clear that the problem of whether economics or politics exerts a greater influence over agrarian structure needs to be more carefully articulated. For certain aspects of the history of agrarian structure, such as the way in which the concept of property is configured, demographic and economic developments were of central importance. For other aspects, such as the precise ways in which feudal rights were redeemed, political forces were crucial. Thus, the nineteenth century offers both concrete information and some analytical insights. This state of affairs adds new fuel to our historical query: how did the relationship between agrarian structure and society express itself in formal terms, during the earlier period, and how regional configurations within the Alps come into being?

Endnotes

1 "Dans l'ensemble, on peut dire que la montagne alpine est, précocement, une démocratie de petits propriétaires, qui détiennent également des biens collectifs"; "un esprit très poussé d'individualiste et de sentiment de proprieté, pour ce qu'il a en propre et, d'autre part, un goût très vif pour la gestion des affaires communes, économiques, puis politiques" (Paul Guichonnet, "Le développement démographique et économique des régions alpines," in *Le Alpi* 1974–1975, 2: 138–96, citation from 2: 149–50). Perry Anderson citation from *Lineages of the Absolutist State* (London, 1979), 301. In the French, Italian, and Swiss scholarship one finds a series of similar declarations, while Austrian scholars reach different conclusions (see n. 15).

2 In the literature on this topic the concept is used in different ways; see Werner Conze, "Agrarverfassung," in *Handwörterbuch der Sozialwissenschaften* (Stuttgart, 1956), 1: 105–13; Heide Wunder,

Chapter Six

"Agriculture and Agrarian Society," in *Germany. A New Social and Economic History*, ed. Sheilagh Ogilvie and Bob Scribner (London, 1996), 2: 63–99, here 74; for the effort to develop a generally-usable typology, see Hans P. Binswanger et al., "Power, Distortions, Revolt and Reform in Agricultural Land Relations," *Handbook of Development Economics*, ed. Jere Behrman and T.N. Srinivasan (Amsterdam, 1995), 3: 2659–2772.

3   This theme is treated, for example, by Viazzo 1989, esp. 263–68 (with respect to inheritance, with specific references); Norbert Ortmayr, *Heirat und Familie in den europäischen Alpen*, unpublished ms., University of Minnesota, 1990 (concerned mainly with illegitimacy, with statistical evidence); other studies will be cited below.

4   Controlled regression analysis is an important method for comparative studies on agrarian structure; part of the historical literature takes its cue from the late nineteenth-century and early twentieth-century situation, but fails to point this out or even to be aware of it.

5   See above all F-Statistics 1862, cxiii; Sandgruber 1978, 122–29; I-Statistics 1930, vol. 2, pt. 1, 1–10.

6   In order to compare two indicators (illegitimacy, occupation) from two points in time, the districts from the 1869 census are taken as a baseline.

7   See the introduction to the statistics that are provided in table A.2.

8   In the Alpes-de-Haute-Provence, where individual property is generally very frequent, half of the area is indicated as "superficie non cultivée." When compared to other departments, the percentage of these lands in farms of 5–10 hectares and of 10–40 hectares was especially high (F-Statistics 1892, 211–25).

9   A-Statistics 1902, xv; the best region for comparison would be the Trentino, which was not included in these statistics. The canton of Ticino was chosen here as a markedly small-farming region (CH-Statistics 1905, 75*).

10 The total number of workers per farm increased in Carinthia from 1.8 (for farms of 0–0.5 hectares) to 8.8 (for farms of 50 or more hectares); the percentage of agricultural servants among these workers grew from 4% to 59% (A-Statistics 1902, 31).

11 Gaston Imbert, *A la découverte d'une population, étude démographique des Alpes-Maritimes* (Aix-en-Provence, 1958), 21; Édouard Baratier et al., *Atlas historique Provence, Comtat, Orange, Nice, Monaco* (Paris, 1969), map 198–99; Émile Levasseur, *La population française. Histoire de la population avant 1789 et démographie de la France comparée à celle des autres nations au XIXe siècle* (Paris, 1889–92), 2: 30–39.

12 See no. 17.

13 Siegfried von Strakosch, *Die Grundlagen der Agrarwirtschaft in Österreich. Eine handels- und produktionspolitische Untersuchung* (Vienna, 1916), 153 (quantitative indications concerning the level of agglomeration or fragmentation of property for the territories of the Hapsburg crown, based on official data collection from the 1870s). The variation of these configurations diminishes the economic comparability of the data, but increases their social significance.

14 The distinctive traits of the eastern regions are consistent with the percentage of medium-sized and large farms (over 30% throughout those regions); estimates are partially interpolated due to missing data.

15 For the "Austrian" perspective, see Mitterauer 1986.

16 Sources such as table A.1. in the appendix; in the Austrian census of 1869, for example, there was no category for agricultural workers who were also family members. They were sometimes

recorded as "permanent servants" and at other times as "persons without a definite occupation." Likewise, agricultural servants who were not family members were imprecisely recorded in the statistics. The map of Mitterauer 1986, 191, based indirectly on the census, is inaccurate. See also Gustav Adolf Schimmer, "Die unehelich Geborenen in Oesterreich 1831–1874," in *Statistische Monatsschrift* 2 (1876): 149–74, esp. 164; *Knechte. Autobiographische Dokumente und sozialhistorische Skizzen*, ed. Norbert Ortmayr (Vienna, 1992), 297. That family members were officially classified as servants is itself of social-historical significance.

17   Michael Mitterauer, "Familienformen und Illegimität in ländlichen Gebieten Österreichs," *Archiv für Sozialgeschichte* 19 (1979): 123–88, citation from 180; id., "Auswirkungen der Agrarrevolution auf die bäuerliche Familienstruktur in Österreich," in *Historische Familienforschung*, ed. Michael Mitterauer and Reinhard Sieder (Frankfurt a.M., 1982), 141–70; Mitterauer 1990, 233–87. For long-term developments concerning agricultural servants, see also Peter Schmidtbauer, "The Changing Household: Austrian Household Structure from the Seventeenth to the Early Twentieth Century," in *Family Forms in Historic Europe*, ed. Richard Wall et al. (Cambridge, 1983), 347–78, here 357.

18   Georges Lefebvre, *La grande peur de 1789* (Paris, 1932), 49; Nicolas 1989, 66.

19   Lefebvre 1932 (see n. 18), esp. 47–50, 197–201, 212–15; Nicolas 1989, chapters 2–3 (also discusses the role of the Dauphiné during the first phase of the Revolution); see also Baratier 1978, 2: 397–438; Bligny 1973, 323–37; *Histoire de la France rurale*, ed. Georges Duby and Armand Wallon (Paris, 1976), 3: 19–28.

20   Ibid., 3: 28–37; Blum 1978, esp. 389.

21   The most complete overview of agrarian reforms is found in Blum 1978, including all of continental Europe, from the Pyrenees to the Urals, except for Italy; a collection of descriptions of the situation in various countries, some of which employ dated research approaches, is in *L'abolition de la féodalité dans le monde occidental. Actes du Colloque, Toulouse 12–16 novembre 1968*, 2 vols. (Paris, 1971); for Savoy see Blum 1978, 83–113.

22   From the perspective of the history of ideas, see Dieter Schwab, "Eigentum," in *Geschichtliche Grundbegriffe. Historisches Lexikon zur politisch-sozialen Sprache in Deutschland* (Stuttgart, 1975), 2: 83–113.

23   See for example *The New Cambridge Modern History*, vol. 9 (Cambridge, 1965), 250–74; maps are located in vol. 14 (1970): 44–48.

24   Blum 1978, esp. 367, 422.

25   William E. Rappard, *Le facteur économique dans l'avènement de la démocratie moderne en Suisse. L'agriculture à la fin de l'Ancien Régime* (Geneva, 1912), 132–204; Hans Brugger, *Die schweizerische Landwirtschaft in der ersten Hälfte des 19. Jahrhunderts* (Frauenfeld, 1978), 367. The effective abolition in Switzerland over the course of the nineteenth century has barely been studied.

26   Mathieu 1992, 82–86; Stephan Franscini, *Neue Statistik der Schweiz. Nachtrag, aus der italienischen Handschrift übersetzt* (Berne, 1851), 105.

27   Jerome Blum, *Noble Landowners and Agriculture in Austria, 1815–1848. A Study on the Origins of the Peasant Emancipation of 1848* (Baltimore, 1948). Friedrich Lügte distances himself from Blum on some issues; see "Die Grundentlastung in der Steiermark," *Zeitschrift für Agrargeschichte und Agrarsoziologie* 16 (1968): 190–209; also Roman Sandgruber, *Ökonomie und Politik. Österreichische Wirtschaftsgeschichte vom Mittelalter bis zur Gegenwart* (Vienna, 1995), 215–17, 233–36 (also for what follows).

Chapter Six

28   Karl Grünberg, "Die Grundentlastung," in *Geschichte der österreichischen Land- und Forstwirtschaft und ihrer Industrien 1848–1898* (Vienna, 1899), 1: 1–80, citation from 49.

29   National figures according to ibid., 70–80; the critique of Lütge (see n. 27) is not crucial for our approximate regional comparison; Stolz 1949, 389–400, citation from 394 (the total burden per cultivated area indicated here should be integrated with other, better-suited indicators that point to a lighter burden); for differences within the Tyrol see ibid., 395–96 (redemption totals) and the sources cited above, chapter 2.

30   "Serfdom" became a hot-button topic during the eighteenth century; see Renate Blickle, "Leibeigenschaft. Versuch über Zeitgenossenschaft in Wissenschaft und Wirklichkeit, durchgeführt am Beispiel Altbayerns," in Peters 1995, 53–79.

31   Various prerequisites for agrarian reform in northern Italy will be indicated below, chapter 8, n. 9.

32   Blum 1978, 102–3, 118, 436; Hugo Penz, *Das Trentino. Entwicklung und räumliche Differenzierung der Bevölkerung und Wirtschaft Welschtirols* (Innsbruck, 1984), 169–81. Rental contract arrangements in around 1900 can be sifted, albeit imprecisely, from statistics relating to farm establishments; see F-Statistics 1892, Modes d'exploitation; CH-Statistics 1905, table 13; A-Statistics 1902, table 12.

33   *Code civil des Français* (Paris, 1804), art. 745 (citation), 832 (against fragmentation); and in general, "Code Law Systems," in *International Encyclopedia of the Social Sciences* (New York, 1972), 9: 214–17; M. Tcherkinsky, "The Evolution of the System of Succession to Landed Property in Europe," *Monthly Bulletin of Agricultural Economics and Sociology* 32 (1941): 165–95.

34   Ulrich von Mohr, *Geordnete Gesetzes-Sammlung und grundsätzliche Uebersichten der Achtzehn Erbrechte des Eidgenössischen Standes Graubünden nebst einem Entwurf zu einem allgemeinen Erbrechte für den ganzen Canton* (Chur, 1831); *Bündnerisches Civilgesetzbuch. Mit Erläuterungen von P.C. Planta* (Chur, 1863); Arnold Niederer, "Bäuerliches Erbrecht. Kommentar," in *Atlas der schweizerischen Volkskunde*, pt. 1, fasc. 7 (Basel, 1968), 570–600.

35   *Geschichte der österreichischen Land- und Forstwirtschaft und ihrer Industrien 1848–1898* (Vienna, 1899), vol. 1, esp. 23–27, 282–331, 343–44, 468–87; Ingrid Kretschmer and Josef Piegler, "Bäuerliches Erbrecht. Kommentar," *Österreichischer Volkskundeatlas*, fasc. 2 (Graz, 1965), 1–18.

36   See chapter 8, map 8.1 and n. 22.

37   Paolo Grossi, *Un altro modo di possedere. L'emersione di forme alternative di proprietà alla coscienza giuridica postunitaria* (Milan, 1977), esp. 194, 237, 320.

38   Albera 1995, chapter 20; Nadine Vivier, "Une question délaissée: les biens communaux aux XVIIIe et XIXe siècles," *Revue historique* (July–September 1993): 143–60, here 145–55; *Geschichte der österreichischen Land- und Forstwirtschaft* (see n. 35), esp. 81–134; *Atlas der Republik Österreich*, fasc. 5, pt. 2 (Vienna, 1972), map 8/3 (Wald: Besitzverhältnisse).

39   See Blum 1978, 434.

40   See n. 16.

7 Territories during the Early Modern Period

From the end of the Middle Ages until the French Revolution, agrarian structures shifted less rapidly, generally speaking, than they did from the late eighteenth century until the end of the nineteenth century. Nonetheless, the early modern period witnessed numerous changes and, moreover, a great variety of political structures. In order to try to grasp the whole range of early modern developments that occurred within the Alpine space, this chapter will look at three regional examples from different parts of the mountain system: Savoy in the west, the Grisons in the center, and Carinthia in the east. The duchy of Savoy corresponded roughly to today's French departments of Savoie and Haute-Savoie, and belonged to the Savoyard-Piedmontese state formation that stretched across the Alps into the northern Italian plain. The Grisons, often called the Freestate of the Three Leagues, was a relatively independent polity that encompassed, in addition to the existing Swiss canton, also a bordering subject region—what is now the Italian province of Sondrio. The duchy of Carinthia, a hereditary land of the house of Hapsburg, provided the name for the federal *Land* in current-day Austria situated in the same area, on a slightly smaller territory. In terms of surface area, these three territories all measured between 10,000 and 11,000 square kilometers. Savoy had by far the highest population level, though Carinthia experienced the greatest

Chapter Seven

rates of growth during the early modern period. The development of political structures was very different in each of these three Alpine regions over the course of these centuries—a fact that makes for very fruitful comparative study.[1]

For many of the themes discussed here we must turn to the different perspectives and traditions represented in the scholarly literature of each region, along with their respective strengths and weaknesses. Practically the only Alpine-wide studies that exist to date are those dealing with domestic units and families, although (or perhaps even because) historical-anthropological research on the family has only developed as a self-standing field within the past few decades. In this context, the most important study is the recent work by the anthropologist Dionigi Albera on *L'organisation domestique dans l'espace alpin: Équilibres écologiques, effets de frontières, transformations historiques* (Household organization in the Alpine space: Ecological equilibria, the impact of borders, historical transformations). This book presents the author's historically-oriented research, carried out in the Alpine valleys of Piedmont, mainly in the Val Varàita. Then it provides an overview of local and regional family studies from many parts of the Alps, on the southern side from Piedmont to Friuli and Slovenia, and on the western and northern sides from Provence to the Valais and all the way to the Austrian *Länder*.[2]

Albera gives particular attention to methodological and theoretical issues. He criticizes research tendencies that portray the family as autonomous and objectified, and argues for a relational approach that sees it as a network of relationships within a constantly changing social context. In this way he de-emphasizes the value of family typologies and positions himself against efforts to construct a European cartography of simplistic family types. His study shows, in detail, that abstract approaches such as this one have generated unrealistic and contradictory representations of an Alpine space already saddled with clichés. The various family types identified by Albera for the study area are thus situated on a medium level of generalization and are explicitly defined as ideal types, that is, as tools of orientation, rather than as immediately realistic descriptions. In my view, these conceptual instruments thus provide a useful introduction to the broad scope of early modern family structures and domestic units. If we synthesize the key points, these types can be described as follows:

> *Agnatic type*: insertion of the family in kinship and neighborhood networks of small settlements; prominent position of the masculine line (agnatic); egalitarian partible inheritance for sons, daughters compensated with dowries;

patrilocal residence followed often by co-residence of married brothers; boundaries of house and family not specified, but constantly interacting with surrounding environment and migrating members. Examples from the Italian Alps.

Bourgeois type: insertion of the family into the organization of the commune, which often includes larger settlements; egalitarian partible inheritance among all children; bilateral kinship; modest social stratification; public roles for men linked not to household management, but above all to local civic rights (bourgeois). Examples from the Valais.

Bauer type: social relations focused on the farm establishment and its operator (*Bauer*, farmer); the commune consists in the totality of clearly bounded, often isolated, farm complexes; forms of public life take household heads as points of reference; undivided inheritance of farm leading to loss of social status for other family members; *Bauer*'s authority based on close ties to lord and to state. Examples from the Südtirol/Alto Adige, Austria, and Slovenia.[3]

As one can see, the typology is not based on the isolation of single elements within the overall household structures, but on the social constellations to which specific familial and domestic relationships correspond. Conceptually we could refer to these as open (agnatic), horizontal (bourgeois) and vertical (*Bauer*) systems. The first two types are related, while the third is more sharply distinctive. The decided contextualization of Albera's synthesis makes it especially useful for historical reflection. In my view, its weakness is located in the uneven and haphazard way in which the relationship between family and environment is treated within each type.[4] The following pages try to clarify this issue (as well as others) by examining the history of three regional states during the early modern period. We begin in the west, with Savoy.

Savoy: the duke, the notables

On 2–3 April 1559, at Cateau-Cambrésis, a city on the border between the kingdom of France and the Spanish Low Countries, signatures were affixed to peace treaties and marriage contracts of great importance for the duchy of Savoy. Emanuel Filibert of Savoy (1528–1580) obtained a sister of the reigning king of France for his wife, but she was already advanced in age and thought to be unlikely to produce heirs. Many of the territories that the house of Savoy had long possessed, but that had been seized by

Chapter Seven

France in 1536, were subsequently restored to the duke, formally as part of Margaret's dowry, according to traditional scholarship. These territories stretched from Burgundy to Piedmont, forming a cushion between the centers of French and Spanish power on either side of the Alps. The peace treaty brought a temporary end to the struggle between the two royal houses for supremacy in Italy.

But slight changes in the international balance of power sufficed to place the duchy in jeopardy once again. In fact, by the end of the eighteenth century Savoy had been occupied militarily by its more powerful neighbors (especially the French), for relatively long periods, another six times. These occupations primarily concerned the territory known as the *Duché de Savoie* (see map 7.1). Beyond this, the dynasty also possessed the duchy of Aosta, the county of Nice, and above all the important principality of Piedmont. Additionally, in the sixteenth century, its dominion extended into regions north and west of Savoy. At the same time as its influence north of the Alps began to weaken over time, it grew stronger in Italy. The southeasterly momentum of this developing state was indicated by the relocation of its capital from the city of Chambéry, in Savoy, to Turin, where Emanuel Filibert made a solemn entry and established his residence in 1563. Contrary to earlier speculation in political circles, the ducal couple had already produced a male heir, and the dynasty's good genealogical luck remained constant. The most famous male successor was Victor Amadeus II (1666–1732), who obtained the royal title that the house had so long pursued. Taking advantage of its geopolitical position, Savoy-Piedmont, or the kingdom of Sardinia as it was now called, had meanwhile become a middling power in European affairs.[5]

This would not have been possible if these fairly non-cohesive territories, beginning with the dynastic restoration of the sixteenth century, had not undergone profound internal changes that transformed them into a centralized state (according to contemporary parameters), with its center at the Turin court. The historiography has not reached a consensus on the precise rhythms and patterns through which this state evolved, but the broad outlines are known for the *Duché de Savoie*, the territory that interests us, located for the most part in the Alps.

In 1561 the last assembly of the three estates was held in Chambéry. Up to that point, this traditional representative institution had been indispensable for the approval of new taxes; thereafter, the duke was no longer interested in discussions and negotiations with this body. Instead, fiscal impositions were now issued by unilateral decree and regularly collected, at least in theory. Specific powers of political oversight were transferred to the high court of justice in Savoy, the Senate, whose original form had been established during the French occupation. The Senate could not veto ducal edicts, but it could delay their application with discussions and objections, though even this became increasingly difficult due to frequent supplementary injunctions arriving from Turin. The Savoyard fiscal court in Chambéry, an outgrowth of the ruling house's patrimonial administration, was able to extend its own influence over the institutionalization of the state financial system, even after a second *Chambre des Comptes*, which soon became more important, was added in the dynasty's capital city. Eventually, though, the Savoyard *Chambre* fell victim to centralization; in 1720 it was abolished, or transferred to Turin. This elimination was linked to the creation of a new political and administrative structure: the intendance. The first intendant-generals for Savoy had already been installed in the seventeenth century, and became a stable arm of the government from 1713 onward, when they were bolstered by provincial intendants in each of Savoy's six (later seven) districts.[6]

Many of the institutional changes introduced beginning with Emanuel Filibert referred directly to the state's fiscal and military interests. When confronted with the imperious behavior of the ministers from Piedmont and the absolutist pretensions of

Map 7.1: The duchy of Savoy in the late eighteenth century

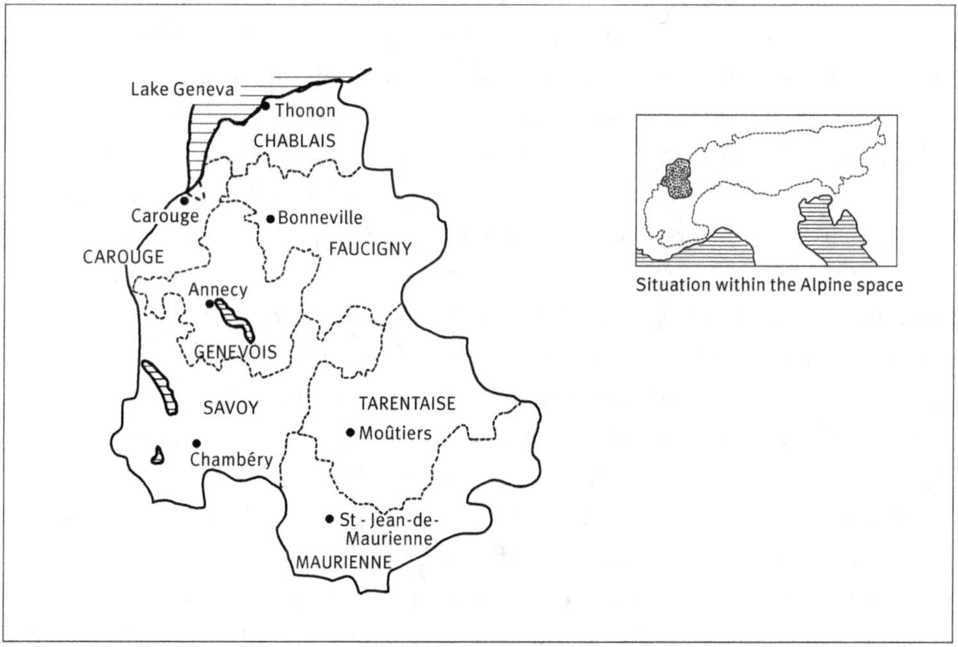

Situation within the Alpine space

the rulers ("our authority is despotic," wrote Victor Amadeus II), one can easily be led to overestimate the power of the center. In fact, the state struggled to establish itself at the local level, as the military events themselves demonstrate.

Emanuel Filibert attempted to create an efficient, commune-based militia for the defense of his lands. In exchange for various privileges, members of the militia recruited from the population at large were to have been transformed into "ducal servants," and thus made directly dependent on the ruler, free from the jurisdiction of local and regional authorities. Following notable initial successes—in 1570 there were an estimated 12,000 militiamen enrolled in Savoy—the institution rapidly lost popularity and subsequently demonstrated its military inefficiency. One reason was the change in the privilege-granting policy, which adapted itself to growing political consolidation and to the increasing control of intermediate powers. Still, over the long term, even the development of communities fell under state influence, particularly since they served as the basic unit of taxation, thus rendering their financial situations matters of public interest. State domination of communities reached its peak with the edict of 15 September 1738, issued in tandem with a new fiscal directive. This legislation established as the formal authority in each commune a small council, whose members were to be recruited through the self-nomination of the wealthiest and most capable residents, while any sort of general communal assembly was prohibited. More influential than the council and its presiding *syndic* (who served a one-year term) was without a doubt the *secrétaire*, a licensed notary who professionally administered a series of communities, and whose nomination had to be approved by the intendant.[7]

The fiscal edict of 1738 was a key moment in Savoyard history. It was based on information collected from the cadastre, or tax survey, which had been initiated ten years earlier. The cadastre mapped each parcel of land in the entire duchy on a scale of 1:2400 and listed the owners' identities. No other civil assignment had required this kind and this amount of effort from state administrators. The *cadastre savoyard* or *sarde* (the dukes of Savoy had become kings of Sardinia in 1720) created a sensation near and far, and, as demonstrated by literary evidence from the likes of Adam Smith, was held to be a masterpiece of land surveying and state administration. Italian surveyors were hired to carry out the project on the technical level, notaries from the countryside compiled the written descriptions of property owners, and the overall organization was in the hands of a specifically-created state office. In the communes, which were supposed to pay for the portions of the project relating to them, there were mixed feelings about the cadastre. "In it, each person could find material proof of his property," writes Jean Nicolas. Inhabitants gave this mapping project credit, he continues, for establishing evidence of ownership, but the ongoing registration effort, which opened property sales and the transmission of land through inheritance to public scrutiny, also alarmed many.

The state's main purpose in this endeavor was to develop more balanced regional and communal tax quotas, which had remained unchanged since first being established in the sixteenth century—this had created significant damage to fiscal justice and thus also to popular willingness to contribute. The question of how to distribute the tax burden, especially the issue of noble and ecclesiastical taxation, had preoccupied all of the dukes and produced large numbers of edicts. The fact that old problems were now being addressed with partially new systems was the result of southern influence: the cadastral registration of the mountainous Savoyard area was an efficient continuation of cadastral mapping in Piedmont and in bordering Lombardy.[8]

What the Turinese government was not able to accomplish in this operation was the registration of the *Servis Ecclésiastiques ou Féodaux*, that is, the rights of tithes and tributes held by the church and the nobility on many pieces of land. For sure the nobility did not collect revenues equal to those paid to the state fisc, but lordly tributary privileges were far-reaching and in some cases very significant. These could include personal servitude with rights over the lands of peasant subjects who died without direct heirs. At the time of the cadastre, the area's nobility constituted a small group, a kind of caste that defended its own privileges tenaciously. Until the mid-sixteenth century, it had been relatively easy for those whose economic power had purchased them greater social status to have themselves recognized as noble. In later periods, Turin used official recognition as a tool of political control and for revenue enhancement. Numerically, ennoblements declined notably: between 1561 and 1600 there were 28 per decade, followed by 19 per decade from 1601 to 1700 and only 6 per decade from 1701 to 1792. As they became more bound to the state, the nobles no longer approached the ruler as a corporation; instead each one was tied directly to the ruler through unmediated obligations. More than any other institution, the Senate and the *Chambre des Comptes* expressed the interests of the aristocracy. Located in Chambéry, both courts were dominated by the nobility, but they also represented a point of access for magistrates seeking entry to noble circles. In general, the centralizing tendencies of the state, and the splendor of residence towns favored models of urban life. Such models helped account for the unequal distribution of the Savoyard nobility across the region's territory. Already in the early eighteenth century, almost 30% of the greatest families lived in Chambéry and a good 50% in its province (Savoie proper), while there were far fewer nobles in remote mountain areas. Seignoral courts, partly tied to feudal economic privileges, were likewise unevenly distributed.[9]

Most nobles lived on their private property, acquired either recently or earlier, and from other resources. Nevertheless they were far from willing to part with their rights to seignorial revenues. This resulted in protracted tugs-of-war between nobles, peasants, and the duke, even in mountain areas. Already in the early seventeenth century, plans were made in Turin to redeem feudal rights that contradicted the state's claims of domination claims. Even before that, Emanuel Filibert had already issued edicts providing for the freeing of his subjects from personal servitude. Until the eighteenth century, though, the issue remained unresolved, since the nobility was ultimately one of the duchy's chief pillars of support. Once the cadastral survey was completed and a movement to free both persons and communes from servitude was initiated, noble privileges fell one after another: on 20 January 1762 an edict outlined the redemption of all rights over persons, and on 19 December 1771 another saw to the redemption of all seignorial economic rights.

We know that the December 1771 edict, which became famous, did not produce the rapid consequences that were hoped for (see chapter 6). One reason for its slow effect was the seignorial reaction that appeared whenever efforts were made to eliminate and modernize property relations. The cadastre had already prompted the nobility to pay greater attention to its own rights, adapting to the circumstances by compiling new registers of feudal tributes. The court was only too familiar with such reactions, but turned instead once more toward the situation in Piedmont, where property structures had long before lost their "attributs féodaux essentiels" [basic feudal characteristics], in the words of one expert.[10]

What can we learn about economic and familial relations within the peasantry? One practically bottomless source of information is provided by the numerous notarial acts recorded in early modern Savoy. Even common people made recourse to notaries for any kind of occasion, business-related or otherwise, such as credit agreements, real estate transactions, marriage agreements, wills, and so on. For example, women's dowry property listed in the acts provides information about economic stratification. The value of such property could reach 1500 florins for wealthy peasants in the seventeenth century, while for middling peasants the average was around 400–500 florins. Most had to be satisfied with around 200 florins, and sometimes even 150 or less (in noble circles an average dowry was held to be 8000–12,000 florins, and a rich one 20,000–50,000).

Whether these amounts were actually materialized depended on gender-specific inheritance practices. As a rule, women were promised dowries from their own families, usually in cash, often to be dispensed in installments. Conversely, care was taken to name masculine family members as principal heirs, in more or less equal portions. Wills could also contain detailed instructions concerning substitutions among the *héritiers universels* (universal heirs). For example, the two sons of the testator are the principal heirs; if one of the two died without leaving male heirs, the other would inherit his portion; if both died without male heirs, the line of the male brother of the testator would inherit. Sometimes wills also contained clauses whose effect must have been the collaborative exploitation of the family property among the principal heirs, for example by way of an inheritance reduction for the party that pressed for the property's subdivision.[11]

David J. Siddle has drawn attention to the importance of such forms of family cohesion in his detailed research on the community of Montmin, located above the Lake of Annecy at more than 1000 meters in altitude, and which was divided into a number of hamlets, like many other localities in Savoy. The farms of these small settlements were not isolated complexes; property parcels were fragmented and distributed across fairly extensive areas. Within households, there were criss-crossing kin relations, and often the inheritance of a previous *chef de famille* remained undivided among the family group. Very few of the eighteenth-century records that have survived refer to divided inheritances; usually they describe property divisions among cousins or between uncles and nephews rather than between brothers. This cohesion within families and kin groups was narrow and patrilineal, and while it did not preclude conflicts, it was able to respond to them flexibly and, in some circumstances, could lead to long periods of economic collaboration. In 1561, 16% of the households in Montmin were composed of multiple families (for example, the families of two married brothers) and 22% included other relatives in addition to the nucleus of parents and children (a nephew, for example). This distribution could naturally vary considerably over the course of the year, but available statistics show that the cohabitation for multiple families for extended periods was a widespread phenomenon during the early modern era. In 1561 and in 1778, in the nearby commune of Chevaline, multiple families accounted for 12% and 20% of households (see table 7.1, p. 178). Elsewhere similar numbers can be identified.[12]

Chapter Seven

In this model it is not difficult to identify affinities with the agnatic family type defined at the beginning of this chapter and described by Dionigi Albera on the basis of evidence from the Alpine regions of Piedmont and other parts of northern Italy (for quantitative similarities, see the Piedmontese examples in table 7.1: Alagna, Pontechianale, Casteldelfino). His investigation is squarely opposed to juridical interpretations of family relations in the southern and western Alps that see property transactions among kin as quasi-automatic executions of fixed customary or written laws. Along with other practice-oriented researchers, Albera shows how wills and marriage contracts reveal the many different, often situationally-specific, individual interests that motivated members of the rural population.[13] His observation seems perfectly true to me, nonetheless, in my opinion, family relations in Savoy remain incomprehensible if one completely ignores the legal order—understood not so much in terms of individual rules or statutes, but with reference to the social history of the juridical field.

Savoy, like most of the southern and western Alps, belonged to the areas in which a written legal culture, which had emerged mainly in Italian cities, had spread during the later Middle Ages. Writing had first been monopolized by clerics, and then became a tool manipulated by men of the law who called themselves notaries and obtained approval from authorities, such that their documents began to be accepted as official. In Savoy during the second half of the thirteenth century, lords began to try to regulate notaries' activities, which underwent a remarkable level of development during the late medieval period and became an important social force during the early modern era. Over the course of the state-building process, jurisdictional activity also grew more differentiated and professionalized. In the fifteenth century, the house of Savoy created new judges with broader territorial jurisdictions, and in the sixteenth century the Senate became a high court for the entire duchy. By this time, even seignorial courts featured professional judges, though over the long term they lost influence within the juridical and state-related public space that was being constructed.

In around 1700 there were about one thousand legal professionals ("hommes de loi") in Savoy. They constituted the bulk of the group typically referred to as notables. The most prestigious among them were university-trained lawyers with ambitions to secure official positions, often in state service. But most of this group were notaries who were distributed across the territory, even in small localities, and who acquired considerable influence by operating at the intersection of economics and the law; family life and public life; seigniory, administration, and local population: "Intermediaries between landowners and debtors, lords and peasants, intendants and local residents, they held a key position in the social edifice of the province."[14] Notaries drew their incomes from a whole range of activities, sometimes drawing up only a few acts each year. Beginning in 1697, Turin required them to consign two copies of their registers at a newly created state office. Simultaneously, limits were placed on the how many new notary positions could be sold by the state. Until the French Revolution, it appears that the number of notaries remained fairly constant at a high level: in 1789, in the Tarentaise, a very mountainous province, there were three offices in which acts could be registered, and in the main office 33 notaries from at least eight localities deposited 2640 acts, the majority of which dealt with the usual economic issues.[15]

Why did inhabitants want to have property transactions among relatives and friends notarized? There is no easy answer to this question. The introduction by Victor Amadeus II of laws recognizing the validity only of wills that were notarized in the presence of seven witnesses, was certainly among the least important reasons why people turned to notaries, since the general tradition had been established centuries earlier. A better question is why families, and especially their male heads, would *not* have wanted to take advantage of this tool of juridical authority. It is obvious that the *chefs de famille*, by means of their flexible alliance with power brokers, increased their own decision-making authority. It would be a mistake to speak of a kind of testamentary absolutism, but such documents definitely carried weight in ongoing and future discussions about the family economy. Gender-related issues of property transmission were also linked to decision-making dynamics. Thus, if on the one hand peasant tactics require our attention, on the other hand it must be emphasized that these choices were made by means of particular instruments, within a particular social frame. This frame was not the same everywhere, as we are about to see.

The Grisons: communes with subjects

In the area that today constitutes the canton of the Grisons, there existed during the early modern period the Freestate of the Three Leagues, a kind of communal republic that dominated a subject territory half again as large, but equally as populated, bordering its southern valleys (see map 7.2). The travelers who visited the republic of the Grisons took pains to describe its social organization, especially in the eighteenth century. One Englishman wrote that it was difficult to recognize a noble there, since depending on circumstances one might even be found running a tavern. A German writer noted that the mountainous country was swarming with noblemen, because anyone who wanted could adorn his name with reference to castles. Differentiation

Chapter Seven

was also missing between cities and their hinterlands: "Chur is held to be the capital city of the Grisons—at any rate, it is the only city," opined a French geographer shortly before 1800, when Chur numbered about 2500 inhabitants and the other two places with municipal rights were smaller than certain villages.[16]

The marked localism, or rather, the minimal level of centralization in the Grisons was a hallmark of its political system that helps explain why European observers were so disoriented there. About fifty large communes, 'jurisdictional communes' in the historiography, had the right to send delegates to the annual Diet. This assembly rotated its annual meeting places among Chur, Ilanz, and Davos; that is, in the territories of the each of the three leagues with which the communes were historically and politically associated: the League of the House of God, the Grey League, and the League of the Ten Jurisdictions. Important decisions were made by delegates to the Diet, according to specific instructions that they brought with them, but the decisions then had to be approved by referendum in each jurisdictional commune. Among the few official appointments that the Freestate, as such, had to offer, by far the most important were those in the subject territory of the Valtellina and the counties of Chiavenna and Bormio, made for two-year terms. Regular taxes were imposed only within this subject region, which was detached from the Grisons' dominion in 1797 and united to the newly created Cisalpine Republic, under French influence. The Grisons, which became part of Switzerland in 1803, only began to develop a more solid state administration over the course of the nineteenth century. Direct cantonal taxes, for example, were introduced in 1856. A cadastre that covered the entire territory was drawn up after 1912, and was completed for all of the communes only at the start of the twenty-first century.[17]

Under these circumstances, there seems to be no other way to describe the eighteenth-century Grisons state than as markedly archaic. In Savoy, regular taxation of the population at large had begun during the sixteenth century and, as described above, the *cadastre* had been created between 1728 and 1738. There are still good reasons not to conceptualize the formulation of state structures as a mono-dimensional process, though. Direct comparison between single factors—such as a cadastre—is problematic, since the function of such factors can change significantly over time. Under the spell of states like France, England, or Prussia, European historiography has long been oriented toward unilinear models which, beyond their size and military strength, were prominent in terms of their pretensions to centralized administration and domination. More recent studies give greater weight to actual practice and to the diverse forms of state development. "The development of a few states in western Europe into centralized bureaucratic and absolutist monarchies," one reads in a review essay of recent research, "is inadequate as a standard of reference for developments throughout Europe."[18] In our case, we find that a localized state such as the Grisons allowed for developments in specific areas that were introduced on a wide scale much later in centralized states like Savoy. This was the case above all for the elimination of feudal seignorial relations, for which little pressure was needed in the Grisons.

At the end of the Middle Ages, dynastic territorial formations were not well developed here. The dominion of the bishop of Chur was the most complex and had established its pre-eminence among feudal seigniories. Alongside this network of aristocratic lordships, there were two other great federations, one formed of nobles and communes and the other of communes alone. The practical and formal relations among these three associations, later referred to as "Leagues," had developed slowly, over time, into a shared polity. Still, given this model of diffused authority, the impetus for state consolidation had to come from without, for the most part. The subsequent unfolding of events was

Map 7.2: The Freestate of the Three Leagues before 1797

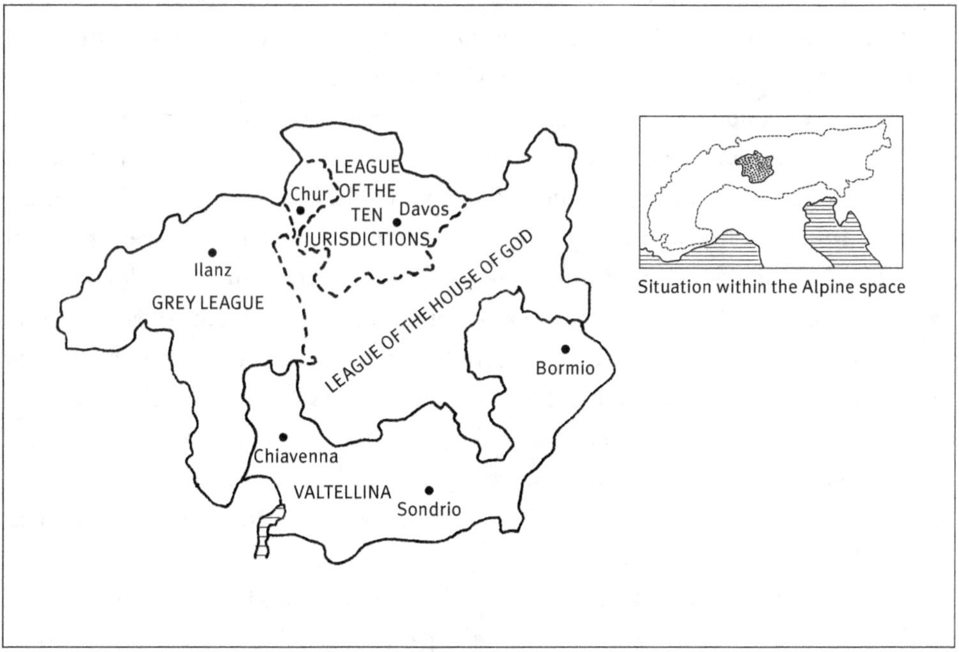

Situation within the Alpine space

Chapter Seven

conditioned by the fact that these outside powers were not particularly interested in seizing control of this territory, and tended to neutralize each other's influence over the Leagues. In order to resist the spread of sovereign Hapsburg power from the east, the Leagues turned to the French crown—as did their neighbors, the Swiss cantons to the north. More external pressure for consolidation came when the Leagues conquered the subject territory on the southern slope of the Alps described above, in the course of the Italian wars. From 1512 onward, shared lordship over these formerly Milanese lands was largely responsible for the institutional construction of the Freestate.[19]

Already during this period, an important role was played within the territory by the representatives of the communes, and this role grew stronger during the 1520s under the influence of the German Peasants' War and the evangelical movement. In 1526, delegates at the Diet issued a document of twenty articles that massively limited the bishop's dominion and attacked many forms of dependence and traditional obligations. The bishop was excluded from future Diets and lost his civil authority, even though he continued to play a political role. His weak position encouraged the introduction of new religious orientations and practices, which in this general trend towards confessionalization led to the creation of an autonomous Reformed church. Significantly, decisions for or against the Protestant Reform were taken up in the first instance by individual jurisdictions and communes. This was the source of a confessional fragmentation within which the Reformed side eventually formed a majority, the Catholic parishes accounting for a good third of the total.

The local situation was also particularly important for changes in the economic elements of the structure of lordship. The 1526 articles decreed cuts in tithe levels and strengthened the position of the peasantry in tribute relations, establishing the complete inheritability of property by male and female children. However, this shift took place only as the result of direct pressure that varied from one valley and one locality to another. Because of widespread disorder, in subsequent decades there were repeated episodes of refusal to pay tributes and a rapid elimination of tithes and dues by means of cash compensations. Together with these events, but not usually simultaneously, an ever-increasing number of jurisdictional rights, which at that time remained in the hands of the feudal aristocracy, were purchased by jurisdictional communes. In this way, official appointments within these localized territories acquired a much more public character.[20]

Some seignorial property and jurisdictional rights remained in effect in the Grisons until the nineteenth century, but as far as the general exercise of power is concerned these had lost considerable importance beginning with the wave of liberations and abolitions during the sixteenth century. The elite increasingly, and later almost exclusively, turned for support to the ownership of widely dispersed private property, credit operations, and income from mercenary service and public office-holding. The new look of the elite also had social and political dimensions. It was typical for nobles in the early modern Grisons to have strong links to their localities, but to lack general institutional backing. In order to exercise wider political influence and to have access to offices in the subject territories, wealthy families had to hold predominant positions within the single jurisdictions and communes that had the right to make appointments. In order to achieve such a position, one had to create a local following of peasants by means of patron-client relations. The nobility differentiated itself from lower social orders through dynastic family policies centered on male lines of descent and consequent matrimonial alliances. Within these tight webs of kinship a specifically aristocratic culture developed. But since the Diet issued no official confirmation of social rank, noble status within domestic politics always remained quite subjective. In 1598, for example, one author referred to about 170 "excellent and laudable houses," and another account soon thereafter included 400 eminent families, while a source from 1666 mentioned only just over 20 great households. The simple fact that such lists were compiled is evidence of the inclination to formalize and reify status—a tendency that began with the elite but involved significant portions of the population during the seventeenth century, when emphasis on titles and the acquisition of new ones became a widespread phenomenon.[21]

Just as personal status eluded the control of any central authority, so was it impossible to impose any standardized territorial subdivision across the entire country. The Freestate was composed of a network of bodies with different rules for exercising, coordinating, or submitting to authority. Over the long term, intermediate powers lost their importance, giving a more prominent profile both to local communities and to overall state institutions.

At the lower levels, this trend—which lasted until the nineteenth century (when it even accelerated)—was partially a consequence of the decentralization of existing associations, driven by competition among communes for greater corporate status. Thus, in the League of the Ten Jurisdictions between 1613 and 1679, three jurisdictions were divided, creating smaller yet more numerous offices, such that communes that previously held a sixth or a fifth of the jurisdictional rights now held a third or a half. Especially striking was the new prominence acquired by local ecclesiastical associations, which had long served as points of reference for communal identities. Over

Chapter Seven

the course of the early modern period the number of parishes in the Grisons more than doubled. Other aspects of communalization on a lesser scale can be found in the ever more frequent written records of decisions and statutes, in territorial boundary-setting and civic laws, and in the creation of communal budgets with occasional revenue collection. Each commune's assembly of those men holding political rights (for which being the head of one's own household was often unnecessary) was and remained an important source of authority. In communal leadership positions it was important that individual neighborhoods be represented, while electoral procedures themselves varied widely. In some places there were direct elections by the assembly of males with rights, but elsewhere elections were indirect (via electors), or appointments were made by cooptation or by drawing lots.[22]

In opposition to this lower-level decentralization, there was an upper-level movement of state centralization, though this phenomenon really only became apparent after 1750 or so, and assumed an institutional form only after the end of the Old Regime. The fact that legislative and jurisdictional activity remained essentially located at lower political levels was a noteworthy obstacle to the process of state unification, and was reflected in the non-professional character of judicial practices in the Grisons. Even voluntary kinds of judicial activities, such as notarial work, were so undeveloped in so many areas that nineteenth-century state legislative reforms were unable to build upon them. This undifferentiated popular culture was echoed by the scarcity of written sources for family law. We mentioned above how long it took to codify inheritance law comprehensively, even though the old statutes were substantially congruent, both in terms of their effectively egalitarian treatment of sons and daughters, and in terms of their strong limitations of the testator's freedom (see chapter 6).[23]

The similarity of early modern inheritance rules came about partly as a result of reciprocal adaptations of local practices in response to conflicts engendered by marriages between persons from different villages. Stabilization of these customs was effected through a pragmatic, case-by-case process. In the Lower Engadine, a valley of three jurisdictions bordering on the Tyrol, the sixteenth-century statutes contained no articles pertaining to inheritance law, while in the seventeenth century a mixture of detailed rules and references to specific cases was established, and was further clarified and amplified by new precedents in the eighteenth century. The widespread use of oral procedures makes it difficult to investigate inheritance practices, but everything points toward the assessment that real, and fairly egalitarian, division of property between siblings, male and female, already represented the usual form of inheritance among the early modern rural population.

As in many other valleys of the Freestate, settlements in the Lower Engadine were mainly comprised of more or less large villages. This facilitated property division, since men's and women's parcels could be combined in various ways to create new farms, through marriages between persons from the same village. In theory this kind of system was much more vulnerable to a process of property fragmentation than was an inheritance system based purely on male privilege. In practice, though, it depended on many factors, especially demographic ones, and did not necessarily result in the atomization of parcels. Much depended on demographic developments as far as farm sizes are concerned as well. Celibacy rates recorded in the Lower Engadine indicate that heirs did not or could not regularly create new households on the basis of their inheritance portions. For example, at Tarasp in 1750, one quarter of the population of persons aged 50 or older were unmarried. Among the women (who rarely emigrated), the percentage was almost a third. The effect of this single population was to increase the number of households that included relatives, protecting family property over the long term.[24]

In table 7.1 I have gathered some data relating to household and family in the Grisons. These show that the percentage of households with extended family members varied considerably, even within the same locality, depending on the demographic situation—at Tarasp, for example, between 13% in 1705 and 27% in 1750. In contrast to what obtained for Savoy-Piedmont, the Grisons witnessed few multiple families (category 5). An important reason for this was undoubtedly to be found in gender-neutral property inheritance, which reinforced the individual claims of sons and daughters and accelerated the division of the family inheritance. The embodiment of familial continuity here was not to be found in the direct economic cohesion of male family members. For this reason, households' growth potential remained modest.[25]

In the southern valleys of the Grisons, developments took a different and quite instructive turn. Unlike the rest of the country, these valleys were influenced by Italian juridical culture during the late medieval period. The first notaries whose presence has been identified came from the area near Como and Milan, and from the thirteenth century onward growing numbers of experts and practitioners of writing and the law appeared. In the early modern period, however, communal regulations effectively subordinated notarial authority to themselves. Within the local jurisdictional system, it appears that independent legal professionals, especially erudite ones, were

Chapter Seven

Table 7.1: Household and family in Savoy-Piedmont and in the Grisons, 1561–1832

Region	Percentages of households with given structure							
Place and year	1	2	3	4	5	1–5	N	Ø
SAVOY-PIEDMONT:								
Montmin 1561	2	3	56	22	16	100	131	5.5
Montmin 1832	4	4	55	29	8	100	124	5.2
Chevaline 1561	0	0	49	39	12	100	41	5.3
Chevaline 1778	0	5	40	35	20	100	20	5.4
Chevaline 1832	7	0	54	25	14	100	28	6.1
Alagna 1734	14	9	43	20	14	100	189	4.7
Alagna 1778	7	7	52	16	18	100	184	4.8
Pontechianale 1826	5	4	56	18	18	100	245	4.7
Casteldelfino 1830	0	6	42	23	29	100	220	5.9
GRISONS:								
Lostallo 1665	14	13	57	14	3	100	95	4.0
Lostallo 1757	7	15	54	20	3	100	59	4.0
Tarasp 1670	4	9	64	19	4	100	69	5.3
Tarasp 1705	14	4	68	13	1	100	79	4.0
Tarasp 1750	6	4	62	27	0	100	77	4.0
Müstair 1762	5	7	71	13	5	100	126	4.5
Alvaneu 1767	5	8	73	11	3	100	73	3.9
Tujetsch 1768	8	8	81	3	0	100	217	4.1
Savognin 1823	6	8	76	6	3	100	62	4.9

1 = single persons; 2 = unmarried brothers and sisters or other unrelated persons; 3 = simple family (married couple and children); 4 = extended family (married couple, children and relatives); 5 = multiple family (two or more married couples whose members were related, and children); N = total number of households; Ø = average household size.

Classification according to *Household and Family in Past Time*, ed. Peter Laslett and Richard Wall (Cambridge, 1972), 28–31. Sources for Savoy-Piedmont: Siddle and Jones 1983 (see n. 12, also personal communication); Viazzo 1989, table 9.1; Albera 1995, chapter 9, tables 1–5. Sources for Grisons: Mathieu, *Storia dei Grigioni* (see n. 16), table 3; Communal archive of Lostallo (parish register 1641–1819); Parish archives of Tarasp (parish register 1610–1716), Müstair (family registers, 18th century), Alvaneu (Status animarum 1763–1859), and Savognin (B 4/5a, excluding St. Martin).

seen as devoid of public utility, if not dangerous and parasitic.[26] As Pio Caroni has demonstrated, inheritance practices in the Italian Grisons that were mediated by the notarial system underwent considerable change. From the sixteenth century onwards, both the freedom of the testator to dispose of his goods as he wished, and masculine privileges in general, were gradually limited. This brought the inheritance practices of this region in line with those of the dominant parts of the consolidating Freestate. In the Poschiavo valley, for example, daughters' renunciations of inheritance portions became less and less frequent, and over the course of the seventeenth century dowries came to represent a regular share of the inheritance.[27] To use Dionigi Albera's terms, this shift can be described as a transition from a more agnatic type to a more bourgeois one. The history of the southern Grisons reveals in broad strokes how family relations that had been molded by particular cultural experiences could be modified by certain political factors. This is worthy of attention, not least because similar processes can probably be identified for other mountainous regions in Switzerland.[28]

At this stage we should stop and compare the two regions from the western and central Alps that we have highlighted so far. To get straight to the most important point: between the end of the Middle Ages and the French Revolution, differences between the Savoyard area and that of the Grisons increased markedly. At the end of the fifteenth century, both regions featured dominions organized by estates, and relatively porous structures of nobility. Three hundred years later, Savoy was a state characterized by ducal centralization, while the Grisons was an exemplary model of communal localism. The reasons for each of these developments are located partly in the specific dynamics linked to the pre-existing historical context of each region, and partly in international influences experienced over the course of the early modern period. The chief differences concern the form and reach of the public domain.

In contrast to what happened in the Grisons, Savoy saw the development of a field of professional legal activity embodied among the notables, and more precisely among the *hommes de loi*, at the same time as the relatively rapid construction of an urban center. In around 1500, Chambéry and Chur each counted 2000–2500 inhabitants, and while the Savoyard capital quadrupled in size over the next two centuries, the population level of Chur stagnated.[29] Key elements of state power and administration were established in Savoy hundreds of years earlier than in the Grisons, but this very fact also contributed to delayed development in other areas. The massive anti-tithe peasant revolt that shook the duchy to its core in 1790 would have been unthinkable in the Grisons during that period, where tithes had long been of marginal importance. While the legislative activity of the Revolution and the *Code civil* of 1804 introduced new prescriptions by which inheritance was to be effected in Savoy "without regard to gender," in the Grisons gender-neutral inheritance had been long established.[30] A bit further below we will revisit in broader terms this issue of the impact of state power

Chapter Seven

on inheritance forms (see chapter 8). Here it will suffice to emphasize an important similarity among the examples considered so far: in both regions the agrarian structure developed primarily around small farmers. Different situations await us in the eastern part of the Alpine space.

Carinthia: Lord, peasant, servant

On September 1, 1557, Augustin Paradeiser, burgrave of Klagenfurt, sent a report on the situation of the peasantry in the duchy of Carinthia to the Hapsburg government in Vienna.[31] If a peasant receives his farm holding from the lord who owns it as a *Freistift* (free, reversible tenancy), wrote this representative of the Carinthian nobility, without specific agreements, either the subject or the lord can cancel the arrangement at the annual village judicial assembly. Even if a peasant and his heirs had held a farm for a hundred years or more, this was no guarantee of ownership. Nonetheless, it was the custom of the country that, upon the death of a peasant, the lord would name as successor one of the peasant's sons, subject to payment of a *Verehrung* (homage or *lods*). This successor had to be a competent manager who knew how to govern his farm, house, and fields well, to pay his rent regularly, and to satisfy other requests made by the lord. In addition to the *Freistift*, the burgrave also mentioned a *Kaufrecht* (purchase right), which gave peasants an *a priori* right to pass on their land through inheritance, especially when this right was demonstrably attested in writing.

From this report and numerous other writings one learns that in Carinthia seignorial rights were widespread in general, but when examined in detail they varied considerably and left wide room for maneuver. Territorial juridical norms were just being created, and at any rate the differences between *Freistift* and *Kaufrecht*, in practice, were sometimes blurred (occasionally one even finds mention of a *Freistiftkaufrecht*). What was important was the individual relationship between lord and subject.[32]

This power relationship was redefined precisely during the mid-sixteenth century—in parallel with rapid state development in the Hapsburg duchy. In the same fashion as their representative in the burgrave's office, many nobles during this period pursued a strategy of requiring a new contract each time a new owner acquired a rural property. The annual confirmation was a ritual act, and many dues were fixed so that the flexible *Verehrung*, collected at the moment when possession passed into other hands, was the best way to increase revenues or at least to recuperate losses caused by inflation. Especially in response to strong demographic pressure or high demand, lords could collect large amounts of money through their right to decide and freely name the person who would take over a farm. Another strategy, though a short-term one, was to sell inheritance rights, that is to collect ahead of time the monetary value of the power of appointment. This strategy was applied most often to princely patrimonial lands, most of which had been mortgaged at the beginning of the early modern period and then sold off altogether. In the years after 1570, subjects of the seigniories that had been mortgaged were solicited to purchase inheritance rights good for two generations. This *Verkaufrechtung* did not catch on among the population, even though the commissioners sent to the countryside for this purpose used every means at their disposal to convince subjects to accept the offer.

The reason for this instant fundraising scheme was the financial squeeze in which the sovereign found himself, and which grew continuously tighter, making itself felt through new taxes and administrative practices involving patrimonial lands across the country. Following the princely model and under his pressure, one seigniory after another began to modernize its own administration. Traditional summary registers listing property parcels and their incomes (*Urbare*) were rewritten with greater accuracy in many places, beginning in the late sixteenth century, and thereafter updated continually via the introduction of "lods books" (*Ehrungsbücher*). When moveable property of subjects was divided, registers were drawn up, ever more frequently, because the lord had a specific interest in the economic situation of the successors, and imposed fees for the losses suffered by moveables taken out of the farm by non-succeeding heirs. The administrative activity carried out by the nobility began to be widely replaced by the work of procurators. In other words, the duchy underwent a massive extension of written documentation and of seignorial control.[33]

Similar observations can also be made with respect to the organization of territory and state formation, which was not very advanced in Carinthia at the start of the early modern period. The Hapsburgs, who held the ducal title since 1335, rarely resided in this hereditary Alpine territory located on the periphery of their sphere of interest, where some of the lords with wide-ranging jurisdictional rights were foreigners. One important impetus for territorial organization came from the war with the Turks. When cavalry units from the Ottoman army and other more or less autonomous groups made repeated armed incursions during the late fifteenth century in Inner Austria, Styria, Carinthia, and Carniola, a far-reaching military and fiscal system that would have a long-term impact on the duchy's evolution began to take root. Owing to the absence of the prince and to the general balance of political power, the regional nobility became the driving force of this process. In 1518 the provincial estates of Carinthia,

Chapter Seven

in which seigniors and knights predominated and the peasants were not even represented, were made collective lords of the city of Klagenfurt by the Hapsburg emperor. The city rapidly became the main regional center, distinguished by its fortifications and a series of monumental buildings (see chapter 4). The provincial estates found in Klagenfurt a fixed site for their meetings, which were usually held once, and then later two or three times, each year. Structures emerged to facilitate the permanent activity of new governmental organs, beginning with the burgrave and a progressively developing administration. Not even the representatives of the sovereign prince, and in the first instance the provincial governor (whose competence included the ability to convoke the provincial estates), were politically and financially independent of this territorially-organized nobility.[34]

A certain centralizing effect derived from the increase in Klagenfurt's profile. New castles began to spring up in the city's surroundings. But elsewhere in the region as well the reconstruction or new establishment of castles and fortified burgs reached a high point in the period after 1560 or so. The Carinthian elites distinguished themselves through the exercise of public authority within their fiefs, where they emulated the lifestyle of rural lords (*"adliges Landleben"*), and in the over fifty judicial districts of the duchy. Jurisdictional authority in seigniories and districts were or could be related to each other in various ways. For example, in 1550 the current provincial governor sold the castle of Sommeregg, which had been the nucleus of an expanding upper Carinthian seigniory and of a judicial district that had recently acquired autonomy. Within this district, Sommeregg extended authority not only over its own peasants, but also over those of other landed lords. On the other hand, Sommeregg's own patrimonial lands also extended beyond the district's clearly delineated borders. Thus, during the second half of the sixteenth century, the new owner made an effort to exchange his lands located in other jurisdictions with lands situated in his district, but belonging to others. By the beginning of the seventeenth century, having expanded as far as politically possible, Sommeregg was a middle-to-large fief in terms of size, with respect to other regional seigniories. What that meant can be gleaned from the mid-eighteenth-century fiscal assessment of Sommeregg carried out by the state: 155 *Huben* (peasant farms), 64 *Zulehen* (agrarian structures managed by other farms), 33 *Keuschen* (smaller farming units—which in Carinthia could still include 2–3 large animals and 5–10 hectares of land).[35]

If the census of farms and smaller agrarian units provided an approximate measure of the dimensions of landed lordship, other measures took revenues into account:

this was the so-called *Herrengült*, which registered the total value of the tributes collected by the seigniory. This revenue estimate, which originated within the financial system of the provincial estates, also became an important criterion for admission to noble status. Until the late sixteenth century it had been relatively easy for wealthy social climbers to break into the provincial elite. An ordinance of 1591 required a specific attestation of nobility in addition to a *Herrengült* of 12 pounds (and soon thereafter, 50 pounds). Contrary to what one might think, however, ennoblements and new entries into the provincial estates did not decrease over the long term, but increased noticeably. During the seventeenth century at least 170 new families or single persons were received into the Carinthian aristocracy (until about 1600, roughly 70 admissions had been registered). This development was made possible in the first place thanks to the sovereign, who during the Counter-reformation and the Thirty Years' War reinforced his effective influence, integrating Carinthia bit by bit into the Hapsburg monarchy by means of ennoblements and patronage politics. 1628 was an especially important moment: by order of Emperor Ferdinand II, the Reformed nobility was given the choice of returning to the Catholic church or leaving Carinthia. This Evangelical-majority region thus became, at least officially, entirely re-Catholicized.[36]

The military sector, which was linked to the financial system, was at the same time an important motor and a central problem of the Hapsburg policy of state integration. The power of the regional aristocracy was based in large part on the fact that new taxes could only be collected through the seigniories, which meant that the prince was forced to make requests and justifications for each new round of impositions, through negotiations that were both arduous and ritualized. Despite this, the sovereign's taxes became permanent during the second half of the sixteenth century. In Carinthia, which was responsible for contributing a certain quota of the amounts owed by the Austrian territories as a whole, the noble estate assembly began to collect back taxes owed, and to hire itinerant tax collectors, beginning in the 1570s. The distribution of the burden within Carinthia was effected in large part on the basis of the landed seigniories and seems to have been done at first through self-declarations. In order to calculate the sums owed by each seigniory—that is, by the peasant farmers therein—the two criteria mentioned above (the seigniory's total revenues and total number of farms) were the chief reference points.

The small- and large-scale interests of the court and the nobility were thus closely inter-correlated in this system. While the sovereign acquired an avenue of fiscal access to the population by way of the provincial estates, the latter could use sovereign authority to increase its own power over the population, to accomplish individual and regional goals. A precise accounting of all the farms spread across the territory was not necessary in order to collect dues and taxes from landed seigniories. The juridical classification as *Huben* or *Keuschen* only reflected the size of the respective agricultural establishments. The use of such broad categories was all the more obvious since many of the farms in the region were isolated establishments with directly contiguous lands; their definition was self-evident.[37]

Just as the prince negotiated taxes with the provincial noble estates, so the nobles or their procurators worked out succession arrangements for the farms under their authority. Through the inheritance tax, or *Verehrung*, which can be documented for many localities beginning in the late sixteenth century, the new peasant proprietor acquired a position for himself, a sort of public office within the power structure. His relations with the lordly authority that certified his possession created a distancing effect with respect to family members excluded from the succession and compensated with moveable goods—these relatives were pushed into a situation resembling that of farm hands.[38] From both the family's side and the lord's side, sons took priority in inheritance considerations, as daughters assumed a secondary role. Daughters still had a chance of being preferred, though, if a wealthy husband were able to offer substantial sums to the lord for the right to be endowed. This assumes, though, that the lord was willing to be swayed by such offers. In extreme cases, the concession of farms could resemble an auction in which anyone could participate. But the general situation can be described as transmission by way of inheritance without inheritance rights.[39]

Depending on circumstances, the possibility of dividing inherited property was also taken into consideration. The mere fact of a landed seigniory's existence did not exclude such arrangements. Given the active economic control exercised by Carinthian lords on their subjects, two parties participated in the decision-making process—the family and the seigniory—increasing from the start the number of hurdles to overcome before subdivision could occur. The facts that many farms constituted territorial complexes that were more or less compact, and that they acquired quasi-official status as units within the evolving tax system, undoubtedly help to account for general hostility against inheritance division in this region. Egalitarian division of farms seems to have occurred less frequently than the new creation of small establishments. On poor, often communal, lands and in growing settlements, *Keuschen* (small farms)

Territories during the Early Modern Period

sprang up. These were often unable to survive as autonomous operations, and had to rely on income from other agricultural or artisanal activities.[40]

This occasional division of farms and formation of minimally-sized units notwithstanding, large peasant establishments were of fundamental importance for the entire early modern period. The assignation of vacant *Zulehen* to well-off peasants contributed to this, balancing the increase in the number of smaller farms by an expansion of larger ones. In order to get a handle on the composition of rural households in Carinthia, family-based classifications such as those employed for Savoy and the Grisons make little sense. A distinctive characteristic of the Hapsburg duchy was precisely the fact that many households included large numbers of non-family members and were stratified in terms of economic status.

Table 7.2 provides a series of quantitative indications from the Gurktaler Alps during the second half of the eighteenth century. The examples show that most of the households in this area included male and female employees. In one parish, an astounding 87% of the establishments had agricultural servants. The renters or *Inwohner* (called *Gästleute* in Carinthia)—who rented an accommodation on the farms and were employed as day laborers during labor-intensive periods—were distributed somewhat less widely. For the most part these were persons who used to be agricultural servants and who, because of age or due to the birth of children, were shifted to this freer yet more precarious position. Not completely integrated into the farm, they were nonetheless juridically and economically dependent on the *Hauswirt*, the head of the household. In contrast to the agricultural servants and temporary laborers, the farmer-owners had life-long rights to the establishment and thereby a support system for old age. In particular, upon the death of the husband or wife, they do not seem to have been unwilling to transmit rights to the farm ahead of time, by means of a formal retirement and consequent assurances of economic support. In any case, analysis of our sample group shows that up to 17% of the households could, at a given moment in time, include retired members. On average, farm establishments counted a great number of persons and thus a large labor force. Not counting children twelve years old and younger, only a minority of the establishments numbered fewer than four persons able (theoretically) to work. As the table shows, households of this size almost always constituted a minority in Carinthia, whereas in the Grisons, where small peasant farms were characteristic, they were a clear majority.[41]

Whether or not establishments included male and female agricultural servants was a key indicator of difference between Alpine regions and between individual farms.

In 1757, in our sample study of the Gurktaler Alps, the average household size was 2.7 servants, while some large units could number up to 10, 15, or even 25. During this period, such hired hands accounted for between a quarter and a third of the population, in the Gurktal and elsewhere in the country.[42]

According to a decree of 1756 regarding rural employees in Carinthia, a worker was to perform his duties "in constant obedience to the will of his lord," diligently and without objections—submission to the head of the household, who had the right to carry out punishments, was a distinctive feature of rural servitude. As Therese Meyer shows in a detailed study, service contracts during the early modern period were usually valid for a year at a time, during which period employees were required to engage in any sort of work on an unlimited basis, whenever needed. When the contract ended, they were free to leave and take a new job elsewhere, and available sources indicate that many of them did just that. Mobility within regions, whether large or small, seems to have been part of the lifestyle of these agrarian service-workers. Their average length of farm employment was between one and two years. If they came from the farmer class, more specifically if they were non-succeeding heirs, they had a higher status and could hope to remain servants only for a certain number of years. In Carinthia, however, many employees were born into the lower order and were servants until they died, whether well-paid grooms or badly paid chicken coop watchers. Frequently they came into the world as out-of-wedlock births. Already for the seventeenth century, the data show that for some years there were localities with illegitimacy rates of 25% and above. Carinthian employees occasionally married each other, though such unions were unusual and tended to lack stability. Jurisprudence on issues relating to agricultural workers began to develop during the late Middle Ages, crystallizing in 1577, when a *Policeyordnung* was issued for Carinthia. Then, from the mid-eighteenth century onward, under Maria Theresa and Emperor Joseph II, more and more detailed regulations were issued.[43]

The overall evolution of the agrarian structure accelerated during this reforming period. While in the sixteenth century the tax policies of the sovereign princes had played a key role in reinforcing and modernizing the landed seigniories, the symbiotic relationship between the princely state and noble states grew increasingly problematic. In the eyes of the Viennese court, international power politics required a real, effective integration of the Austrian lands; that is, direct control over the emperor's subjects, without mediation by landed lords. The comprehensive strengthening of

Table 7.2: Household and economic status in Carinthia and in the Grisons, 1750–1798

Region Place and year	Percentages of households with given structure					
	AS	RE	PE	LP	N	Ø
CARINTHIA:						
Feistritz 1757	87	37	4	6	133	8.2
Gradenegg 1757	57	38	5	31	42	7.5
Liemberg 1757	52	33	4	48	27	7.3
Obermühlbach 1757	75	52	8	18	231	9.5
St. Lorenzen i. d. R. 1757	77	55	10	22	96	9.6
Sirnitz 1757	64	39	6	32	196	7.4
Zweinitz 1757	55	29	16	31	89	7.1
Zweinitz 1770	62	41	9	32	93	7.2
Zweinitz 1786	47	39	17	40	94	6.5
Zweinitz 1798	50	39	17	29	94	6.8
GRISONS:						
Tarasp 1750	0	0	3	61	77	4.0
Lostallo 1757	2	0	2	73	59	4.0
Müstair 1762	3	0	2	70	126	4.5
Alvaneu 1767	8	0	1	74	73	3.9
Tujetsch 1768	0	0	2	—	217	4.1

AS = with agricultural servants; RE = with renters, mostly also employed in temporary agricultural work; PE = with life pensioners (persons who had formally retired from directing the household); LP = with low labor potential (fewer than 4 workers at least 13 years old); N = total number of households; Ø = average household size.

Sources: Vienna Database on European Family History, Institut für Wirtschafts- und Sozialgeschichte, University of Vienna; for the Grisons see table 7.1 (there, 'renters' are an unknown category, and the parent who is not the first person listed for the household is held to be a life pensioner; for Tujetsch there are no age indications).

Hapsburg fiscal authority is demonstrated by the development of the cadastre. In around 1750, Theresian fiscal regulation brought about a general, state-wide tax survey, in the tradition of the sixteenth-century seignorial registers, which distinguished more clearly between peasant and noble property, so that both could be taxed, yet in different ways. Property was measured for the first time by the Josephine cadastre of 1787–88, which sought to apportion property taxes more fairly and to restrict the lord's portion of the agricultural product. The haste with which this project was carried out made mapping impossible. Only in the nineteenth century would a cartographic cadastral survey, with detailed representations of parcels' yields, become politically

Chapter Seven

and organizationally feasible. In the duchy of Carinthia, data collection was effected between 1826 and 1829.[44]

Alongside reforms in fiscal and other sectors, a state administration that functioned independently of the seignorial system came into being. In 1748 Carinthia was sub-divided into three *Kreise* (districts) and then in 1782 these were consolidated into two and furnished with corresponding district offices (see map 7.3). In a parallel development, the communes emerged from their embeddedness in the seigniories, and gradually became the basic politico-territorial units for military, fiscal, and cadastral purposes.[45]

Still, even at the end of the eighteenth century, sums owed to the state could not be collected without the cooperation of seignorial administrators. This created difficulties for the increasingly important differentiation between (public) taxes and (private) revenues, and was reflected in the absence of a clear juridical definition of property. Precisely who held authority over soil and land in the widespread *Freistift* tenure arrangement was especially unclear, so sorting this out became one chief activity of reformers. Beginning in 1784 a series of sovereign edicts transformed all

Map 7.3: The duchy of Carinthia in the late eighteenth century

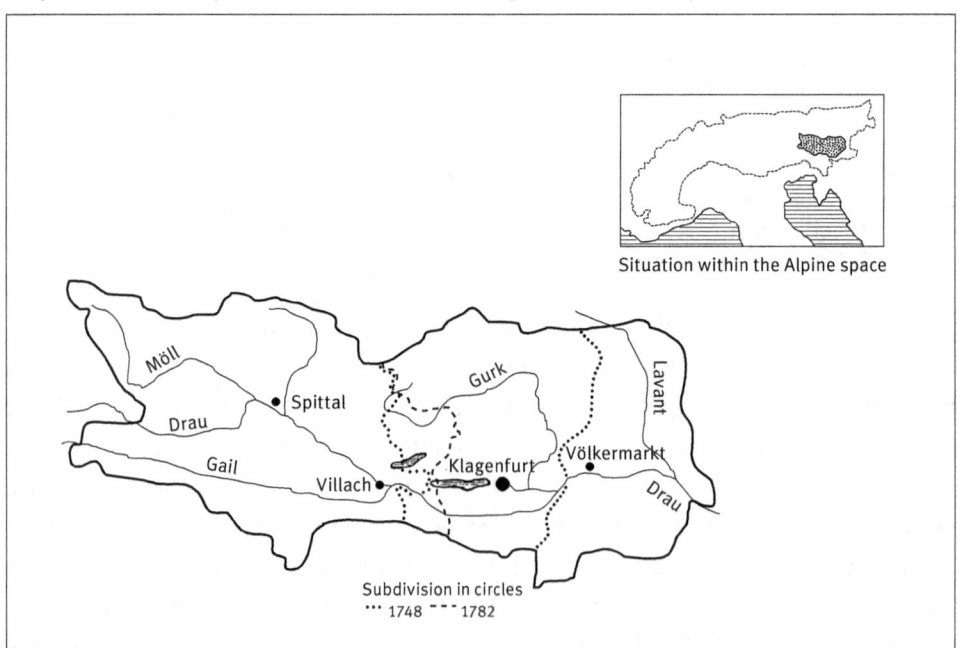

Freistift relations into *Kaufrecht* or *Erbrecht*, thus giving owners the right to sell or transmit their rural property. The lods were not thereby abolished, but transformed into fees imposed on property exchanges, just like those collected through *Kaufrecht*. The amount of this fee was regulated, though, independently of the social control that lords could individually exert to the advantage or disadvantage of their subjects. Nobles who criticized the reform movement argued that the "gentleness and forbearance" of the landed lords would decline as a result, since they had been deprived of the freedom to "increase, as the need arose, their *Ehrungen* revenues." In this way, the authority of the sovereign prince had damaged each landed lord's "direct dominion."[46] Indeed, this was in the first instance a matter of the relationship between the monarchy and the regional nobility. The *Bauern*, whose economic situation did not change in the short term, displayed little interest in this discussion. The "direct dominion," the unmediated lordly authority over land—as opposed to the *Bauer*'s use right to the land—was hindered, but the sub-division of the property concept continued to exist through *Kaufrecht* and the various traditional tribute payments. Thus, the situation created in the duchy of Carinthia in 1784 was roughly the same as that whose abolition had begun thirteen years earlier in the duchy of Savoy.

In light of these regional and political developments in the western, central, and eastern parts of the Alpine space, we are able to establish the following provisional assertions:

(1) Early modern family and household conditions in the three regions can definitely be described according to the typology offered by Dionigi Albera for the entire Alpine arc. Our overview shows that in Savoy and the Grisons, respectively, "agnatic" and "bourgeois" family models predominated, with the real sub-division of property (masculine in the former case and gender-neutral in the latter); while Carinthia offered an example of the *Bauer* type, with the undivided transmission of property.

Our analysis draws particular attention to two issues. First, a certain correlation could be found between household structure and settlement patterns. In Savoy and especially in the Grisons, the aggregation level of settlements was clearly higher than in Carinthia, which was characterized by isolated large farms. Second, families were situated in particular ways within different kinds of sociopolitical contexts. In the Grisons, inheritance was regulated for the most part without recourse to public authorities, and regulatory norms were really general guidelines offered up by gradually developing communal statutes. In Savoy, heads of households established flexible partnerships with the bearers of notarial legal culture and authority—via the composition of

Chapter Seven

written wills—whereas in Carinthia they exercised a quasi-official function in their own right, through public alliances with local power-holders.

(2) Processes of territorialization and state formation in the three regions were marked by the influence of social forces whose dimensions varied. The growing importance of the sovereign prince and the communes in early modern Savoy and the Grisons, respectively, led in turn to centralized or localized structures, while in Carinthia the seignorial nobility emerged as an intermediate authority. These different power distributions were reflected in the diverse chronologies of constitutional change in each region—not that some regions were more progressive or traditional than others, but rather that the development process in each region was selective and path-dependent.

This lack of developmental simultaneity becomes apparent when one compares state and seignorial revenue collection and cadastral mapping in each area. In Savoy state taxation and country-wide cadastral surveys began centuries earlier than in the Grisons, which, for its part, was the first area to eliminate (or rather, to individualize or communalize) seignorial dues. "Lordly claims" survived until a particularly late date in Carinthia, where princely fiscal impositions did not take place on an independent basis as they did in Savoy, but were mediated by the nobility. This substantially reinforced the public influence of the landed seigniories and was one reason for the quasi-official character of the *Bauern*; it also accounts for the impression of "feudalism" associated with this eastern land.

(3) On a wider level, there are also surprising similarities among the three regions. In each of them, there was a rapid consolidation of territorial state institutions during the sixteenth century, mainly after 1550, whose chief characteristics proved resilient to change until the end of the Old Regime. If the absence of simultaneity mentioned above resulted from different internal political arrangements, then this sixteenth-century simultaneity can be said to stem from the external power struggles of the time, along with the military and fiscal innovations that were drawing vast parts of Europe together into a *Konfliktgemeinschaft* (community of conflict). The unfolding of these regional histories seems to have been remarkably influenced by the internal political configuration at the beginning of the modern period. Existing differences could be amplified through political consolidation, and thus prefigure later trajectories. Conceptualizing such a model of path-dependent development obviously requires a broader foundation. Whether and how this model can be generalized is one of the themes of the next chapter.

Territories during the Early Modern Period

Endnotes

1 The territory of Savoy was altered many times, and was estimated to cover an area of 10,800 km² during the eighteenth century; the area of the other two regions was about 10,300 km² each. For the percentage of Alpine territory, the average altitude of communal centers, and the best available population estimates, see chapter 2, tables 2.2 and 2.3.

2 Albera 1995; also id., "Familles. Destins. Destinations. Entre mosaïque et portrait-robot," *Le monde alpin et rhodanien* (1994): 7–26; Pier Paolo Viazzo and Dionigi Albera, "The Peasant Family in Northern Italy, 1750–1930: A Reassessment," *Journal of Family History* 15 (1990): 461–482; the Austrian studies of Michael Mitterauer are especially important, see Mitterauer 1990; a comparison with situations in Switzerland is found in Jon Mathieu, "Von der verstreuten Familie zum 'Ganzen Haus'. Sozialgeschichtliche Übergänge im schweizerisch-österreichischen Alpenraum des 17. bis 19. Jahrhunderts," in *Der Vinschgau und seine Nachbarräume*, ed. Rainer Loose (Bolzano/Bozen, 1993), 245–255.

3 Albera 1995, chapters 7, 12, 16; in ibid., chapter 18 there is also a subvariant of the agnatic ideal type; unlike the other two types, the agnatic one is not represented in the text by a fixed nomenclature.

4 For the *Bauer* type Albera stresses the significance of the political environment, but then he explains the agnatic type only in terms of its economic context (an open economy oriented toward emigration). For the bourgeois type his study seizes cursorily upon a formula (horizontal alliances of rural communes against outside powers) that works badly with his relational approach; on this issue see below, chapter 8.

5 Important treatments of Savoy-Piedmont are Roger Devos and Bernard Grosperrin, *La Savoie de la Réforme à la Révolution française* (Rennes, 1985); *Il Piemonte sabaudo. Stato e territori in età moderna*, vol. 8/1 of *Storia d'Italia*, ed. Giuseppe Galasso (Turin, 1994); Guichonnet 1996; Nicolas 1978; Guido Quazza, *Le riforme in Piemonte nella prima metà del Settecento*, 2 vols. (Modena, 1957); "Savoia," in *Enciclopedia italiana di scienze, lettere ed arti*, vol. 30 (Rome, 1936), 925–55; Enrico Stumpo, *Finanza e stato moderno nel Piemonte del Seicento* (Rome, 1979); Geoffrey Symcox, *Victor Amadeus II. Absolutism in the Savoyard State 1675–1730* (London, 1983). The themes discussed here are often dealt with in many studies; I will provide citations to particularly pertinent passages and for additional specialized studies.

6 Devos and Grosperrin 1985 (see n. 5), 32–34, 47–51, 412, 418, 422; Nicolas 1978, 2: 596–615.

7 With his famous sentence, mentioned here in abbreviated form, Victor Amadeus II made the claim that his authority in Savoy was not limited by any corporate body. However, it was understood, and not without good reason, in more general terms; see Devos and Grosperrin 1985 (see n. 5), 418; Walter Barberis, "Die Bildung der 'milizia paesana' in Piemont: Zentrale Gewalt und lokale Verhältnisse zwischen dem 16. und 17. Jahrhundert," in *Klientelsysteme im Europa der Frühen Neuzeit*, ed. Antoni Maczak (Munich, 1988), 261–97; Quazza 1957 (see n. 5), 1: 73–76; Nicolas 1978, 2: 616–21.

8 Guichonnet 1996, 224 (for citation of Jean Nicolas); Paul Guichonnet, "Le cadastre savoyard de 1738 et son utilisation pour les recherches d'histoire et de géographie sociales," in *Revue de Géographie Alpine* 43 (1955): 255–98; Max Bruchet, *Notice sur l'ancien cadastre de Savoie* (Annecy, 1977); in general Kain and Baigent 1992.

9 Nicolas 1978, vol. 1, esp. chapters 1 and 4; Guichonnet 1996, 236.

Chapter Seven

10   Jean Nicolas, "La fin du régime seigneurial en Savoie (1771–1792)," in *L'abolition de la féodalité dans le monde occidental* (Paris, 1971), 1: 27–108; Guichonnet 1996 (see n. 5), 225 (for citation).

11   Devos and Grosperrin 1985 (see n. 5), 195, 276; Gabriel Pérouse, "Etude sur les usages et le droit privé en Savoie au milieu du seizième siècle," *Mémoires de l'Académie des Sciences, Belles-Lettres et Arts de Savoie*, 5th ser., 2 (1914): 305–631 (example from 330); Roger Devos, *Vie et traditions populaires savoyardes* (Lyon, 1991), esp. 32–33, 73–74.

12   Siddle 1986; id., Articulating the Grid of Inheritance: The Accumulation and Transmission of Wealth in Peasant Savoy 1561–1792, in Mattmüller 1986, 123–81; id., and Anne M. Jones, "Family Household Structures and Inheritance in Savoy 1561–1975," *Liverpool Papers in Human Geography* 11 (1983); indications for other Savoyard communes in Devos and Grosperrin 1985 (see n. 5), 254.

13   Albera 1994 (see n. 2), esp. 20 which refers to Savoy; also Albera 1995; Laurence Fontaine, "Droit et stratégies: la reproduction des systèmes familiaux dans le Haut-Dauphiné (XVIIe–XVIIIe siècles)," *Annales ESC* (1992): 1259–77; David J. Siddle, "Mediation and the Discourse of Property Transfer in Early Modern Europe," *Rural History* 6 (1995): 11–28 (Savoy).

14   Nicolas 1978, esp. 1: 72–84 (citation from 79), 2: 606–15, 843–57; *La pratique des documents anciens*, ed. Roger Devos et al. (Annecy, 1980), esp. chapters 1–4; François Vermale, *Les classes rurales en Savoie au XVIIIe siècle* (Paris, 1911), 200–45; Jean-Paul Poisson, *Notaires et Société. Travaux d'Histoire et de Sociologie Notariales*, 2 vols. (Paris, 1985/1990) (includes several studies on Savoy); *Problèmes et méthodes d'analyse historique de l'activité notariale (XVe–XIXe siècles)*, ed. Jean L. Laffont (Toulouse, 1991) (for the Briançonnais among other things); Laurence Fontaine, "L'activité notariale (note critique)," *Annales ESC* (1993): 475–483.

15   Poisson 1990 (see n. 14), 94.

16   Silvio Margadant, *Land und Leute Graubündens im Spiegel der Reiseliteratur 1492–1800. Ein Beitrag zur Kulturgeschichte und Volkskunde Graubündens* (Zurich, 1978), 108 (citation), 157, 252. Classic works on the Freestate of the Grisons include Johann Andreas von Sprecher, *Kulturgeschichte der Drei Bünde im 18. Jahrhundert* (Chur, 1976 [1875]); Friedrich Pieth, *Bündnergeschichte* (Chur, 1945); Oskar Vasella, *Geistliche und Bauern. Ausgewählte Aufsätze zu Spätmittelalter und Reformation in Graubünden und seinen Nachbargebieten* (Chur, 1996); for recent discussions of the state of the research, see the *Handbuch der Bündner Geschichte* (Chur, 2000), esp. the articles on the late medieval period by Arno Lanfranchi/Carlo Negretti, Florian Hitz, and Roger Sablonier, and those on the early modern period by Jon Mathieu, Max Hilfiker, Randolph C. Head, Silvio Färber, Guglielmo Scaramellini, Martin Bundi, Ulrich Pfister, and Marc Antoni Nay. References will be given for detailed passages, and to other specialized studies.

17   Sprecher 1976 (see n. 16), 492–95, 704–7; Pieth 1945 (see n. 16), 448–49; personal communication from Erwin Müller, Office of Improvements and Surveys of the canton of Grisons.

18   Practice-oriented studies have demonstrated that the level of centralization, bureaucraticization, and absolutism have often been overestimated in "model states"; see the general editors' preface to *The Origins of the Modern State in Europe, 13th to 18th Centuries*, ed. Wim Blockmans and Jean-Philippe Genet (Oxford, 1996), vol. D, vii.

19   The loss of the subject territory between 1620 and 1639 was linked to the deepest crisis ever experienced by the Freestate; the best synthesis of the late medieval (until 1512) development of the

Grisons is Sablonier 2000 (see n. 16); for the state-building phase, see Head 2000 (see n. 16).

20   Pfister 2000 and Mathieu 2000 (see n. 16); see also on this topic Marc Dosch, "Bündner Gemeindewesen an der Wende zur Neuzeit," *Lizentiat* thesis, University of Zurich, 1996, esp. 36–37, 49–56.

21   Mathieu 2000 (on local clienteles) and Färber 2000 (on language used to refer to nobles) (for both, see n. 16).

22   In Grisons historiography, the commune enjoys a privileged position, but communal dynamics during the early modern period have not been adequately studied; see some indications in Mathieu 2000 (for the increasing density of local organization) and Pfister 2000 (on the church) (for both, see n. 16).

23   *Bündnerisches Civilgesetzbuch. Mit Erläuterungen von P. C. Planta* (Chur, 1863), esp. 454, 457; J. Regi, "Differenzas prinzipalas tanter il vegl ed il nouv dret d'eredità," in *Chalender Ladin* 6 (1916), 69.

24   Jon Mathieu, *Bauern und Bären. Eine Geschichte des Unterengadins von 1650 bis 1800* (Chur, 1987), 41–44, 142–45, 174–77; for the history of settlement patterns, see Mathieu 1992, 66–74, 149–59.

25   In the Grisons, kinship relations played an important role, perhaps increasingly so over the long term, but their configuration tended not to be agnatic and tied to a particular group, but open and bilateral.

26   The scholarship on notaries focuses on the earlier period—see Lanfranchi/Negretti 2000 and Hilfiker 2000 (see n. 16); hostility toward the juridical class was a widespread phenomenon during the early modern period, but was particularly acute in smaller territories, see Pio Caroni, "Statutum et silentium. Viaggio nell'entourage silenzioso del diritto statuario," *Archivio Storico Ticinese* 32 (1995): 129–60.

27   Id., *Einflüsse des deutschen Rechts Graubündens südlich der Alpen* (Cologne and Vienna, 1970), 120–215, here esp. 196–99; the specificity of the southern Grisons can also be identified in the household lists, where from time to time one finds married brothers within the extended co-residential family.

28   This was the case for example in the central and lower Valais, see Albera 1995, chapter 16; Pierre Dubuis, *Le jeu de la vie et de la mort. La population du Valais (XIVe–XVIe s.)* (Lausanne, 1994), esp. 261–63, 279.

29   For Chambéry see table 4.1 above, the figure for 1600 is very probably too low—see Devos and Grosperrin 1985 (see n. 5), 34; for Chur, see Mathieu 1992, 94.

30   Nicolas 1989, 109, 213, 342; also above, chapter 6.

31   The paragraph that follows is based mainly on Jean Bérenger, "Le Alpi orientali. Gli Asburgo e il pericolo turco 1480–1700," in Martinengo 1988, 237–65; Hermann Braumüller, *Geschichte Kärntens* (Klagenfurt, 1949); Peter G.M. Dickson, *Finance and Government under Maria Theresia, 1740–1780*, 2 vols. (Oxford, 1987); Karl Dinklage, *Geschichte der Kärntner Landwirtschaft* (Klagenfurt, 1966); Fräss-Ehrfeld 1984–1994; Walther Fresacher, *Der Bauer in Kärnten*, 3 vols. (Klagenfurt, 1950–1955); Therese Meyer, *Dienstboten in Oberkärnten* (Klagenfurt, 1993); Armin A. Wallas, *Stände und Staat in Innerösterreich im 18. Jahrhundert. Die Auseinandersetzung um die Gerichts- und Verwaltungsorganisation zwischen den Kärntner Landständen und der zentralistischen Reformpolitik Wiens* (Klagenfurt, no date [1987]). I will provide precise references for pertinent passages and other specialized studies.

Chapter Seven

32   Fresacher 1950–55 (see n. 31), here esp. 2: 46–47, 159–161 (citation from 160), and 3: 55, 77 (this study is rich with information, but full of questionable interpretations, in particular because of its old-fashioned approach to "law," which is treated as a fixed category).

33   Fresacher 1950–55 (see n. 31), esp. 2: 42–43, 56, 65–66, 83, and 3: 43, 52, 77–79, 170–186; Fritz Posch, "Die Verkaufrechtung auf den landesfürstlichen Pfandherrschaften Kärntens im 16. Jahrhundert," *Carinthia* 147 (1957): 465–87; Dinklage 1966 (see n. 31), 99–142.

34   Bérenger 1988 (see n. 31); Wallas 1987 (see n. 31), 23–35.

35   Fräss-Ehrfeld 1984–1994, 2: 545–75; Wilhelm Wadl, "Geschichte der Burg und Herrschaft Sommeregg. Ein Überblick," *Carinthia* 179 (1989): 153–68; Dinklage 1966 (see n. 31), 104; Peter Cede, *Die ländliche Siedlung in den niederen Gurktaler Alpen. Kulturlandschaftswandel im Einzelsiedlungsgebiet unter dem Einfluss des Siedlungsrückgangs* (Klagenfurt, 1991), 54.

36   Gustav Adolf v. Metnitz, "Geadelte Bürger in Kärnten," *Carinthia* 155 (1965): 439 and 156 (1966): 187; id., "Adel und Bürgertum in Kärnten. 17. Jahrhundert bis 1848/49," *Carinthia* 158 (1968): 607; Fräss-Ehrfeld 1994 (see n. 31), 690–708.

37   Fräss-Ehrfeld 1984–1994, 2: 32–35, 116–17, 222–24, 252–54, 418–21; Fresacher 1950–55 (see n. 31), 1: 47–49; Herbert Paschinger, *Kärnten. Eine geographische Landeskunde* (Klagenfurt, 1976), 149–50.

38   In an impressive study of Upper Austria that also relies on some problematic concepts, Hermann Rebel calls attention to the integration of peasants into the power structure (Rebel 1983).

39   Fresacher 1950–55 (see n. 31), 2: 55–104; Dinklage 1966 (see n. 31), 124–25.

40   In 1848, 30.7% of all agricultural possessions in Carinthia were entire farms, 19.1% were half farms, 11.6% were quarter farms, and 38.6% were *Keuschen*, the smallest unit; see Othmar Pickl, "Bauer und Boden in Kärnten und Steiermark zwischen 1788 und 1848," *Blätter für Heimatkunde* 55 (1981): 131.

41   For agricultural servants and day laborers, see Meyer 1993 (see n. 31), esp. 151–77; for recipients of life pensions, see Thomas Held, "Rural Retirement Arrangements in Seventeenth- to Nineteenth-century Austria: A Cross-community Analysis," *Journal of Family History* 7 (1982): 227–54; for potential labor power, see Mitterauer 1986, 242–44 (index of the subordinate class, defined in the same way as LP in table 7.2, but with divergent values, because—among other things—the numbers in many parishes referred only to portions of the population).

42   Recorded percentages of agricultural servants, as part of the total population: 32% in the Gurktaler Alps in 1757 (same sample as table 7.2), 27% in the county of Ortenburg in 1763 (Meyer 1993 [see n. 31], 178); compare to 1 % in the Grisons in 1750–1768 (same sample as table 7.2).

43   Meyer 1993 (see n. 31), citation on 112.

44   Kain and Baigent 1992, 191–203; Dinklage 1966 (see n. 31), 173.

45   Wallas 1987 (see n. 31), 75–76; Braumüller 1949 (see n. 31), 315–20, 376.

46   Fresacher 1950–55 (see n. 31), 2: 104–57, citations on 130, 147.

8 State Formation and Society

If certain geographers of an older school had any say in the matter, we would not even be asking about the importance of political factors for the development of agrarian structure and society in the Alpine space. The ready-made answer would surely be that their impact was marginal. In this vein, a synthetic work written during the early stages of Alpine regionalism claims that the various internal developments of different states left the lifestyles of mountain populations untouched: "The vicissitudes of these states had barely any influence on the way of life of their inhabitants, on this mountain civilization whose roots reach deep into prehistory, and which has lasted, without major changes, until the nineteenth century."[1]

The material examined in the foregoing chapters has led us to a different conclusion. Over the course of the early modern period, all of the territories studied here underwent processes of change that qualify without a doubt as major. From their political and social structures to the intimate circle of family and household, these regions were so diverse that it is difficult to speak of an Alpine "way of life." Beginning in the late eighteenth century, the influence of long-range power relationships began to grow at the expense of short-range ones: with clear variations in timing and relevance, the state's elimination of seignorial privileges resulted in the modernization of property

Chapter Eight

rights. But there was no question of a uniform agrarian structure, not even in around 1900. According to statistics now available for the first time, small farm establishments were the rule in the western part of the Alpine space, while eastern parts had a high percentage of large farms, with numerous agricultural employees.

The concept of 'mountain civilization' deployed by authors such as the one cited above is rashly synthetic and delimiting, and thus of little utility for an investigation of the divergent patterns of development in mountain regions. This chapter takes as its starting point precisely the opposite perspective, by placing Alpine agrarian structures within a European perspective. One of the most distinctive characteristics of early modern Europe was the emergence of different patterns of development in East and West. Fernand Braudel drew upon his literary flair as a historian when he wrote of a Western "heart" of the continent, as distinct from the Eastern "peripheral regions." Without being systematically rigid, he pointed out that in the West one found agrarian structures subject to tribute payments in kind or in cash, and densely populated cities with their urban bourgeoisies. In the East he emphasized the growth of the landed aristocracy and an increase in the direct labor services, or *corvées*, performed by subjects. According to Braudel, these two areas of Europe were separated by a line that ran from Hamburg to Venice, or on a Hamburg-Vienna-Venice axis.[2] If this very broad, almost geometric, schema is taken at face value, the line of separation would in the first case run to the east of Innsbruck, bisecting the Alpine arc, and in the second case the entire mountain range would belong to the West, with just a few outlying districts. In either case, this model makes it immediately and literally clear that our set of historical problems should be situated with respect to European regional differences and zones of transition.

If one wishes to draw attention to the particularities of the Alps, it would make more sense to point to their privileged position from the perspective of research strategy, rather than wasting effort on the "civilization" issue. Indeed, in order to contextualize Alpine agrarian structures, one must refer not only to East-West differences, but also to those between regions north and south of the Alpine watershed—as will be done in the next section. Following this roughly sketched and theoretically oriented overview, we will take the discussion to a broader level, considering the significance of politics in the development of rural societies in the Alps.

The European dimension

The classic study of Ester Boserup on the conditions for agrarian growth in preindustrial societies is a stimulating starting point for a general discussion and debate on this topic. Unlike many economic theorists, she treats agrarian structures, or "tenure systems," not as extra-economic factors, but as internal variables linked to demographic and economic processes. But this linkage comes into play only with respect to one specific element: property formation. With regard to the resulting property allocations her model remains open to political interpretations, as can be seen in this brief summary:

Agrarian intensification, monetization and property formation. In agrarian societies, increases in population and land-use intensity are accompanied by a proliferation of property rights. The scarcer land is in a given territory, and the more continuous the exploitation of individual plots by individual families, the more likely that general rights concerning status in the territory will become less valuable than specific individual rights to the soil. At a certain point in time, often a class of feudal lords emerges as a local government, collects payments in kind from resident families, and is able to employ their labor in farming operations run by the public-lordly authority. Increasing monetization resulting from urban growth alters feudal juridical relations and, together with agrarian intensification, plays an important role in bringing about the privatization of agricultural land.

Power relations and property allocations. Depending on the political balance of power, privatization can take various forms. Sometimes feudal lords are weakened with respect to the peasantry; they lose control of their dependents and are only able to transform into private property the farming establishments that they manage directly. In other cases, lords are able to use force in order to deny peasants' rights to the land, becoming themselves the owners of much or all of the land previously subject to feudal jurisdiction, while the peasants sink to the level of tenants and non-owners.[3]

In historical reality, the processes of property formation and allocation that have been distinguished here for analytical purposes were of course parallel phenomena. Nonetheless, as I refer to the situations found north and south of the Alps, I will occasionally limit myself to a single aspect, because from a European perspective, northern Italy is a perfect example of early property formation, while the agrarian structures in the German area are markedly divergent.

The early privatization of landed property in northern Italy is most easily documented through a reading of the erudite juridical language that, since its medieval appearance, had to rely on generalization and simplification. To describe the relationship between lords and peasants, the jurists tended to make a distinction between the "dominium directum"—the immediate ownership-like authority enjoyed by a lord over a piece of land—and the "dominium utile"—the peasant's right to cultivate the same land. Until well into the nineteenth century, this well-known pair of concepts was used in many regions north of the Alps to define, in a formulaic way, different sets of lordly relations. The Austrian law of 1848 for the abolition of servitude and the liberation of rural properties, for example, eliminated the juridical basis for any due owed because of the *"grundherrliches Obereigenthum,"* that is, because of direct lordly dominion.[4]

In contrast, Giorgio Chittolini and Gauro Coppola show in an overview of the constitutional development of northern Italy that there, the concept of a divided property regime was already outmoded, in central areas, by the late Middle Ages. With the expansion of the authority of the communes and other changes, direct dominion was first juridically transformed into a simple easement, and then into private property that comprised both forms of "dominia." The structure of rural power thus acquired a new, in some ways impersonal, face. Local lordly authorities were replaced by the control of cities and their elites. The land, which belonged to or over time was acquired by non-peasant groups, was now farmed by their salaried labor or rented out according to freely-entered short-term arrangements.[5] As the authors point out, this juridical evolution took place against the background of an exceptional set of economic circumstances. Some of the data concerning early modern northern Italy have been mentioned above. Compared to the contemporaneous situation in other areas, these data show, especially for the sixteenth century but also for the following periods, demographic density, intensive agriculture, and advanced urbanization.[6]

Fiscal reforms were generally tied to the modernization of property rights, and were characteristic of the increasing complexity of social organization in northern Italy. The regional Italian states were innovators in fiscal issues; this can be illustrated particularly well for the eighteenth century. In the duchy of Milan, in 1718, only a few years after the Austrian Hapsburgs had taken possession of the territory from their Spanish cousins, a "Giunta di Nuovo Censimento" (New Census Authority) was created. It proceeded to compile a widely admired cartographic cadastre. The data collected included all of the productive and unproductive property of the duchy, and formed the basis of a specific centralized property tax, according to which the burden, regardless of opposition, was no longer assigned to collectivities. Although the initiative had originated in Vienna, in the majority of the lands of the Austrian crown such a reform would have been unthinkable during that period. Confronted

State Formation and Society

with the strong position of the aristocratic order in those territories, according to the experts on cadastres, "a Milanese-type mapped cadastre would have been constitutionally inappropriate and too controversial to contemplate." In the case of Piedmontese fiscal and cadastral reforms, the initiative came from an Italian center—the court in Turin. Begun at the end of the seventeenth century and later influenced by the Milanese project, from 1728 onward the collection of cadastral data extended from Piedmont into the dynasty's lands in Savoy (see chapter 7). The accelerated pace by which innovations in state structures took place in this Alpine duchy is easier to understand when one takes its southern situation and origin into account.[7]

But the development of northern Italy was neither linear, nor territorially uniform. The seventeenth century was a true crisis period, when population levels as a whole stagnated and cities lost inhabitants. Was this economic and demographic crisis an expression of a persistent, sociopolitical "re-feudalization" process, as a thesis that has been discussed widely over the past several decades would have it?[8] Opinions on this issue remain divided, though the central thesis seems to have lost support, since it presupposes the restoration of situations that had been left behind, whereas in reality state-related and social activities were growing more different from, not more similar to, what had gone before. Giorgio Chittolini argues that we should look at this problem from the broader perspective of the territorialization of regional states. Even in regions like Lombardy where, beginning in the Middle Ages, the cities had witnessed especially dynamic political and demographic growth, the next phase of political expansion was based not on a unified constitutional arrangement, but on particular dispositions for communities and local lordships. The move toward territorial expansion was in many ways more significant than the conflict between communal and feudal interests.

Among the peripheral areas that had been shielded from direct seizure by the cities were the Alpine valleys. Within the developing states, these areas often benefited from a great many privileges and communal autonomy. At the same time, though, they were less exposed to the economic and political influences that were leading toward the modernization of property structures in the more central districts. For this reason, precisely in the Alpine regions, emphyteusis and other forms of dependency for indefinite periods were basic elements of the agrarian structure.[9] This very clearly indicated ambivalence, along with the equally well-distinguished and highly differentiated structures of city-states, are the key elements for our topic. From a Swiss perspective, the central Alpine area in around 1800 looked like a region in which the

Chapter Eight

burden of feudalism was light (see chapter 6), whereas from an Italian point of view, it looked like an area in which the number of feudal remnants was quite large.

From an early date, the East-West developmental patterns mentioned above provided a basis for comparative reflections. "One can observe from city-dwellers and peasants in Germany that in some regions, such as Franconia, Swabia, the Rhineland, etc., they generally enjoy the status of free persons, or are required only to perform certain labor services or make cash payments. But in the Brandenburg marches, Pomerania, Lusatia, Moravia, Bohemia, Austria, etc., they live in a kind of servitude in different ways," remarked a *"Handbuch der neuesten Erdbeschreibungen"* (Handbook of the Most Recent Geographic Descriptions) of 1802, referring to earlier authors.[10] From the late nineteenth century onward, similar observations were re-elaborated through the catchy concept of "agrarian dualism." This notion would posit the Elba as the boundary that developed during the early modern period between two different agrarian structures: to the west, *Grundherrschaft*, a kind of lordship based on tributes from peasant farms in cash or in kind; to the east, *Gutsherrschaft*, a kind of lordship based on large estates managed by nobles and worked by *corvée* labor—the forced labor of peasants tied to the land. As with many other threads of international and German historiography concerning East-West relations, the debate on agrarian dualism can not be separated from its political context—which was extremely sensitive during the twentieth century. This is evident in the schematic conclusions that were drawn regarding the populations who lived under these two kinds of lordly arrangements: individual responsibility on one side, and a spirit of subordination on the other.[11]

Today, within a new context, research has in large part moved in a different direction, taking up suggestions of older, differentiated approaches. There is now interest in precise investigations of the many regional examples of mixed agrarian structures, and also of the spaces in which individuals asserted their agency, and this in a broader European perspective. For present purposes it will suffice to look at a group of three selected studies taken from a very diverse field of research, dealing in turn with the state, lordship, and farming establishments.[12]

Europeanization and regionalization are basic issues in the reflection of Charles Tilly on the long-term process of state building. Distancing himself decisively from unilinear models that take single, eventually powerful, states (such as France and England) as the measuring stick for the entire continent, he emphasizes the multiplicity in kind of state formations, and their remarkable capacity for survival. He sees the

State Formation and Society

formation of the state in Europe as having developed along a curve. In an early phase the pre-existing structural differences among individual regions were reinforced, resulting in increased variation, and in the second phase there was an international realignment, resulting in the creation of nation-states that were relatively similar everywhere. Tilly's question is thus "why European states followed such diverse paths but eventually converged on the national state." In order to explain and model this diversity, he distinguishes "capital-intensive" paths from "coercion-intensive" ones. In regions where the capital-intensive model prevailed (areas with many cities, dominated by commerce), state formation was stamped by a socioeconomic context that facilitated the creation of military power while blocking the ability of military specialists to seize control themselves. The opposite occurred in the coercion-intensive regions (predominantly agricultural areas with few cities), where the firm grip on the bureaucracy held by warlike feudal lords, along with the submission of the peasants and urban bourgeoisie, was decisive. The author takes his typical examples from the very area that interests us, referring for his capitalist model to the highly urbanized regions that stretch from northern Italy northward across the upper Rhine, and for his coercive model to eastern regions like Hungary.[13]

There is no reason to follow Tilly in this conflation of modes of production and political power—a mixture derived from traditional bourgeois and Marxist *clichés*—thereby reinforcing questionable elements within the concept of agrarian dualism. The notion that predominantly agricultural regions based on a natural economy were necessarily dominated by powerful nobilities is inadequate, as is the idea that the process of state formation in regions where an urban and monetized economy prevailed was more or less free of coercive practices.[14]

But the following other observations made by Tilly seem important for our general discussion:

(1) When describing the causes of regionally-specific developmental paths, the author points out the facts that earlier developments conditioned later ones; that the ideal objectives of individual powerholders played a limited role in state formation; and that innovations had to be negotiated or imposed within a given context of power relations.

(2) The process of state formation carried forward by competition among the dominant members of society, and was was as such collective; the transition from a non-cohesive cluster of state-like entities to a true system

of states with formalized relations can be dated to the end of the Middle Ages; a push toward this formalization was provided by the Italian wars of the 1490s.

(3) Following the dramatic expansion of the military sector during the early modern period and the growing density of international relations there was, especially from the early eighteenth century onward, a levelling-out process; by this point state formation was determined to a larger degree than before by external forces, and in the course of the nineteenth century a generalized model of national states prevailed; in this way, movements aimed at altering internal balances of power also took on a national character.[15]

Thus, while political developments during the early phases of state formation depended on what happened regionally, thereafter they were influenced by dynamics operating at a much wider scale.

But we are still left with the question of what the regional power configurations on the northern edge of the Alpine space looked like during the earlier period. With the title *Zwischen Ost und West* (Between East and West), Herbert Knittler has recently offered a quantitative study of landed lordship in Lower Austria during the early modern period, with a comparative look at neighboring regions. Lower Austria includes a non-negligeable amount of Alpine territory, but extensive and important parts of it are located in the Danubian area, providing us a useful point of departure.

At the beginning of the early modern era, the revenues of the landed lordships in this country were based mainly on dues paid in cash and kind by their own subjects; according to available sources, *corvée* labor on directly managed lands was not widespread. The expansion of such noble-operated farms, with labor provided by the *corvée*, was, however, characteristic of later developments. In 1563, the Hapsburg prince granted the provincial estates of Lower Austria the right to the "unlimited *Robot*," which meant that lords could require subjects to perform labor services in amounts to be determined by the lords. This was a fact that reinforced their power over subjects who were by now well anchored in the juridical realm. Still, the direct administration of farming establishments by the nobles seems to have expanded more rapidly only after an initial take-off period. In the eastern half of the region, the amount of land cultivated under the direct supervision of nobles, within selected lordships, increased by 30% between 1620/1630 and 1750. A sovereign decree of 1772–73 provides us with a reliable measure of the degree to which *corvée* labor became habitual; it establishes a limit of 104 days per year for the "unlimited *Robot*." Development took place differently in nearby Upper Austria, where in 1597 the *Robot* was fixed by norm at 14 days per year. Nonetheless, the pressure of the landed lords was not lighter in this area, where fees for property sales and other kinds of cash dues underwent a massive increase right at the beginning of the early modern period. Knittler underscores the relationships among these various tributary obligations, since an increase in the *corvée* beyond a certain level caused a decrease in the payments in kind or in cash that had hitherto been demanded.[16]

One of the causes that Knittler identifies for the divergent paths taken by these two regions during the sixteenth century is a variation in settlement patterns, linked to juridical and economic conditions. According to him, Upper Austrians settled in isolated farmhouses or in small hamlets, preventing the creation of strong peasant communes and facilitating lordly control over individual farms. In Lower Austria, on the other hand, most peasants lived in villages that suffered greatly during the late medieval demographic crisis, thus leaving empty lands that could be seized by lords. Population loss also resulted in an increase in cultivated acreage during the crisis that began in the 1620s. The impact of market conjunctures was less uniform, and seems to have led to different results depending on the specifics of each situation.[17]

To conclude, if one takes the *corvée* and the amount of acreage farmed under the direct supervision of the nobility as the key measures, then Lower Austria fits into a vast area stretching from the Baltic, across the Bohemian crown lands, to Styria and extending also into Hungary and Croatia. For Bohemia and Moravia, for example, during the 1770s, the *Robot* limit was 156 days per year, while in Styria (as in Lower Austria) it was 104 days. The percentage of farmland exploited directly by the nobility in these latter two regions was 9–12%, while in Hungary, according to our indications, it was 15–30%. In the Polish and eastern German areas, where sources show a high density of lordly control, *corvées* and noble-managed farming areas were far more extended—so that these territories, according to Knittler, should be set aside. However, he also observed that the agrarian structure of Lower Austria and adjacent areas can not be described as a lordship in the western sense of the term, but was rather an intermediate form, between East and West, in which both labor services and also payments in cash and kind played their roles.[18]

The third category dealt with by the recent research under consideration here, the farming establishment, is much less frequently employed in scholarly comparisons of agrarian structures than the other two. This impedes our discussion in various ways, especially when an insistence on the categories of 'state' and 'lordship' leads to mistaken judgments about their economic basis.[19] Franz Irsigler has written an overview of the large and small farming establishments in western Germany between the thirteenth and eighteenth centuries that shows how the regional differences between farm structures was just as great as differences among forms of lordship. His work attempts to interpret various kinds of sources, collected during the early twentieth century, regarding agrarian structure, from an historical-developmental perspective.

Southwestern Germany was at that time an area in which small or even very small farms were the norm. As such it differed from most of the German territories, and from neighboring Bavaria in particular, where medium- and large-sized establishments prevailed. A similar arrangement could be noted with respect to inheritance patterns of landed peasant property. In the southwest, partible inheritance predominated, while in Bavaria and elsewhere inheritance was undivided. Irsigler sketches a typology of historical economic zones in terms of farm size and related factors. In so doing, he focusses on the late medieval period, drawing a picture that is more characterized by continuity than by change. His findings show that the southwest was already clearly distinguished from Bavaria before the early modern period. For example, in Bavaria one found widespread small settlements and isolated farms, while on the upper Rhine there were many cities, large villages, and an agricultural situation that could be described as intensive, according to the standards of the period. In around 1500, the region between Basel and Mainz was described by travelers as a garden.[20]

David Warren Sabean's local study on Württemberg extends from early modern times into the nineteenth century and sheds much light on the agrarian structures in the southwest, and especially for the issue of property divisions at inheritance. The author stresses the complexity of European inheritance systems, with respect for example to relations between the sexes; to the bequests of property, houses, and tools; to the monetary liquidation of property; and to use of testamentary practices. He confirms the idea that these various systems are better understood as operating along a continuum between an egalitarian pole and a non-egalitarian one, than as a simple opposition between partible and undivided inheritance. Still, when classifying the region that he studies, Sabean tends to start with the criterion of the partibility of inherited land. As an important element of his explanation he looks at the intervention of state and fiscal authorities during the sixteenth and early seventeenth centuries: at that moment, the general inheritance model was fixed, according to the prevailing balance of political forces.

State Formation and Society

In Württemberg, the duke first collected information about local inheritance practices and then, between 1555 and 1610, established a unified set of gender-neutral inheritance law, which was quickly adopted, by and large. "A particular inheritance regulation may have violated the old custom of a particular village, but once in place for a generation or so would become part of the observed rule structure." The absorption of, or adaptation to, such innovation also took place under bureaucratic pressure, because along with the regional legislation local officials were required to compile a current inventory of the individual shares of the inheritance. This uniform and state-imposed division of landed property did not necessarily result in the fragmentation of farming establishments. Tendencies in the direction of small peasant property seem to have existed from a very early date, though it is also true that only the rapid demographic growth beginning in the eighteenth century led to the subsequent further decreases in farm sizes.[21]

At this point we can return to the region that is the focus of our study. The territories that have been mentioned in this overview, situated to the north and east of the Alpine space, are indicated in map 8.1. The map also shows, very schematically, the tendencies toward direct noble farm management, and toward partible or undivided inheritance practices.[22] From the indications provided in this and the two previous chapters, we can conclude that the East-West differences in agrarian structures take largely similar forms in both the plains and the mountains. In both cases, western farming establishments were small or very small; the situation of medium- and large-sized farms in the eastern portion of the Alps paralleled that of Bavaria and the Danubian part of Upper Austria; to a relatively lesser extent, only in a spotty manner, the eastern portion of the mountains also displayed the tendency toward direct noble farm management. On the whole, the Alpine situation thus seems not to constitute a special historical case, but rather fits into European historical patterns.

Politics as a factor of differentiation

The factual convergences between the Alps and surrounding areas that arise from this comparative analysis also offer useful interpretive clues. Ecological interpretations, along with economic ones, based on agrarian structures, thus become less important, while political ones move into the foreground. The following reflections take as a key assumption something that we have seen repeatedly: that the political field underwent a deep shift as the early modern period began, characterized by what has been described as formalization, densification, and juridical solidification. Despite regional particularities, the shift was general.

Chapter Eight

Among the many causes of this acceleration of state formation in the Alpine space, geopolitics also played a role. Especially during the Italian wars of the late fifteenth and early sixteenth centuries, a large number of the conflicts between, and important alliances among, the European powers straddled the Alps. The first map of the region was drawn up, by order of the French monarchy, to deal with the difficulties involved in crossing the Alps: the king wanted to know which passes best facilitated the movement of his army into Italian territory. On the *Carte d'Italie* that was printed soon thereafter, the Alps were represented from the *Mons de Gaule* (mountains of Gaul) to the *S. Godard* (St. Gothard pass in what is now Switzerland); the explanation accompanying the map pointed out that cannons would be most easily transported over the Montgenèvre pass (see illustration 3).[23]

One simple way to understand these changes is by looking at how political growth created differences. As we have seen from previously-mentioned examples both within and outside of the Alpine arc, regional developments diverged during the early modern period. Consider Upper and Lower Austria, two areas in which landed lordships grew increasingly different beginning in the sixteenth century, and—as in Carinthia—simultaneously grew in importance. One can also point to Savoy and the Grisons, where the sovereign prince and the communes, respectively, grew in significance and created structures that were centralized in the former case and localized in the latter. A key point of departure for the various political paths that were followed was the configuration of power relations in around 1500 between the sovereign, the nobility, and the communes. If one force or another was particularly important then, it had a good chance of consolidating and expanding this influence, in order to dominate the other elements during the state-building process. In other words, the political dynamic created differentiations both within regions and between them. At the same time, these forces had differing ranges of interests and thus unequal developmental potentials. The communes had the fewest trump cards to play in this game, and the prince had the most. Over the long term, though, communal formations showed how important they were, while the noble estate, after having grown stronger during the early modern period, eventually folded. The development toward extensive territories organized as national states led by around 1900, almost everywhere, to a separation of status-related rights from economic ones.[24]

A similar, but less prominent, kind of differentiation can also be observed, or inferred for good reason, with respect to farming establishments. In this case, existing settlement patterns were important. Villages, for practical and conceptual reasons

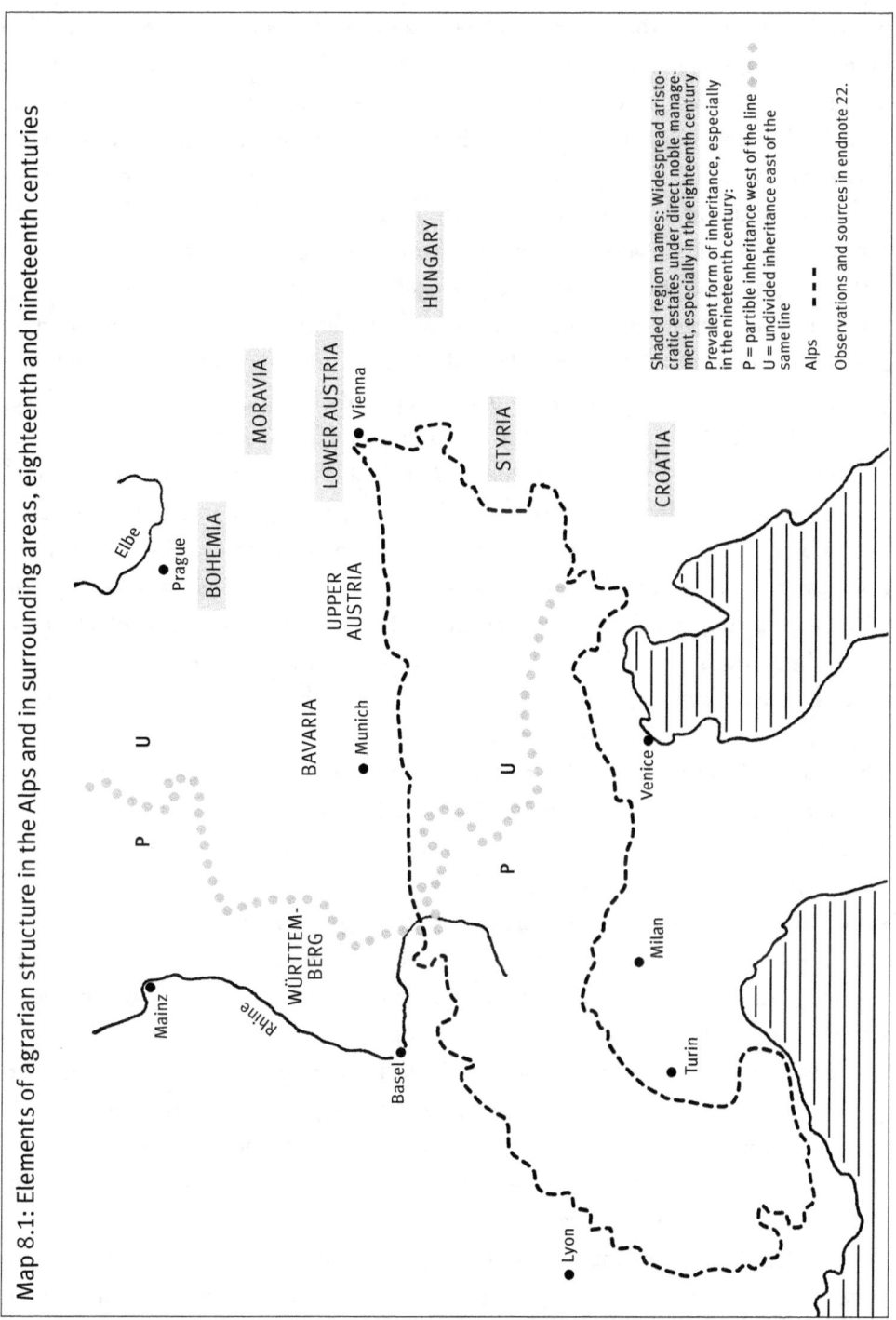

Map 8.1: Elements of agrarian structure in the Alps and in surrounding areas, eighteenth and nineteenth centuries

Chapter Eight

(we will return to this theme), tended toward a mode of production in which partible inheritance and potentially smaller farms were key elements, while in areas of isolated settlements farms could not only expand in size more easily, they also favored undivided inheritance practices. Depending on the settlement situation, this provided a basis for juridical solidification in either direction. But it eventually became clear that this institutional expansion, a product of territorialization and state formation, was an overlay onto local structures, thus contributing to regional differentiation. The social position of the powerholders was of secondary importance. A number of scholars have claimed that the undivided form of inheritance was often imposed by powerful landed lords.[25] In Carinthia, with the strengthening of the landed nobility, passing down a farm was turned into something like the transmission of an office, but only in a limited sense could this be understood as inheritance. And the influence of landed lordship need not come from the nobles, as it did in this case; it could also come from the prince himself. Suppositions regarding the internal relationship between given powerholders and a given structure of farming establishments seem thus less plausible to me than the suggestion that the key to their relationship was chronological: regional developments depended in no small measure on which political configurations were socially dominant, on all levels, at the moment when state expansion took place.[26]

The history of the Tyrol, which, as we have seen, was a transitional region between East and West, is of particular interest for this effort to correlate state structures with household ones.[27] Here, in contrast to the eastern Hapsburg territories, communes with jurisdiction were represented in the provincial diet and directly involved in the system of public finance that was created over the course of the sixteenth century. At that time it also appeared that the degree of institutionalization of inheritance law was increasing, especially in the jurisdictions in which the transmission of property was now being officially certified. On the county level though, one only found general laws regulating property divisions in farms belonging to the landed nobility, and guidelines to follow in the division of peasant farms (territorial ordinances of 1526, 1532, 1573). Given both the wide variation in settlement and inheritance patterns across the Tyrol, and the political influence of the single jurisdictions, a unified approach to inheritance was unthinkable for the time being.

The difficulties of creating such uniformity were revealed during the eighteenth century, when the Hapsburg sovereigns intensified the role of the central administration in this far-flung county, and began to oppose property divisions vigorously. A decree issued by Maria Theresa in 1770 required that the authorities make rulings on the transmission of farms through inheritance, on the basis of specific documents, in order to make sure that the heir receiving the farm would not have to compensate coheirs with cash payments that were too high. In jurisdictions where it was not a usual (or politically enforceable) practice to bequeath farms, divisions had to be avoided as far as possible. The complexity of the situation also manifested itself in the fiscal cadastre of 1775, which registered both farms and agricultural plots. This cadastre subsequently became more important than the legislation itself. Apart from the question of their implementation, the degree to which these laws were valid at all in the southern and western parts of the county was far from clear. Still, the cadastre provided an administrative basis for the regulation of individual farm establishments. Those properties that had been registered under the rubric of a given farm were from that point forward to be considered as belonging to it and no longer divisible. With the legislative anchoring of such *geschlossene Höfe* (closed farms) in 1900, the Tyrol came to be structurally partitioned: divisibility for the non-registered farms, indivisibility for the registered ones. The region's transitional character was now mirrored by its institutional setting.[28]

Rural societies

By way of conclusion we would like to consider some aspects of rural societies that arise, in two different ways, from our preceding analysis. From a theoretical perspective, how do the arguments made so far relate to other explanations of developmental processes and sociopolitical structures? From a historical perspective, what significance do these structures hold for contemporary practice and assumptions? Responding to these questions requires comparison with the existing scholarship and a consideration of other experiences.

There is no reason to reaffirm the opinion that attributes a given, environmentally-determined character to agrarian structures in the Alpine space. It is only necessary to overcome the limited regional perspectives behind such suggestions, to show how far they are from reality. The notion that single elements—such as settlement patterns—are conditioned by the natural environment are clearly invalidated through comparative analysis, too.[29] Models that focus on economic arguments, while using environmental variables as only as background considerations, are much more interesting. In the scholarly literature such models serve to ground the relationships between small and large peasant properties, referring to (1) agricultural technology, (2) type of production, and (3) land-use intensity.

(1) In a study of the forms of social reproduction in rural France during the Old Regime, Bernard Derouet takes a position in favor of greater attention to economic matters, especially agrarian technology, and against an overemphasis of juridical and cultural influences. He argues that specific environmental conditions caused plowing in southern France, for example (in contrast to what took place in the north), to remain technically simple and inefficient, thus favoring small peasant holdings. In these circumstances, small peasant farms did not fall behind the large ones in terms of labor productivity and were better able to preserve their own independence. This is particularly valid for mountain regions, where the economic, ecological, and technical context made it opportune for the peasant farm to limit itself to its own property. The context "also promoted small farms that were all roughly the same size as the best local economic arrangement. Holdings could not be too large since the use of salaried labor was not profitable (due to the techniques employed and to low labor productivity); rather, their dimensions had to correspond to an effective use of the family labor force."[30] This approach raises the question as to whether farm and landholding relations during the Old Regime were determined to a significant degree by labor productivity and by economic rationality understood as such. In general, many indications show that during this period technical factors were substantially less important than during the second half of the nineteenth century, after the revolution in agricultural technology took hold.[31] The example of the eastern Alps proves without a doubt that non-family labor could be "profitable," even in an Alpine context.

(2) Michael Mitterauer deals with this issue of non-family labor in his research on the history of the rural household economy in Austria. "From the perspective of extra labor demands generated by peasant family farms in the area under study, one can make a distinction with respect to ideal types between two principal forms of rural societies: the *Gesindegesellschaften* (society of agricultural servants/hired hands), and the *Taglöhnergesellschaften* (society of day laborers). The first are characterized by an extra labor force that is tied to the family, which receives these workers for at least a year at a time and has them participate in family life; in the second, short-term work relationships, without any substantial integration into the family, are the norm." According to Mitterauer, the key reason for this difference is to be found in the specific kinds of work associated with specific types of production. For example, in wine-producing regions with great seasonal fluctuations in labor requirements, day-labor work makes sense. For

livestock-raising, on the other hand, work is carried out over the course of an entire year, so in these regions hired hands with continuous stable employment arrangements are preferred. The many male and female servants of the Alpine *Hörndlbauern* [horn peasants] are to be explained in the first instance with respect to the type of production that prevails in the area.[32] Indeed, we know that the eastern Alpine regions featured a high number of agricultural hired hands. Still, even livestock-raising was marked by different seasonal phases (summer pasturing, hay-making, stabling), and in the western Alpine regions, where this kind of production was just as important, the numbers of hired workers were low. This suggests that the size of the farm, rather than the seasonality of the type of production, was the decisive factor. If only a small number of animals had to be taken care of, the family labor force sufficed.[33]

(3) One of the reasons offered to explain the subdivision of property and the circumstances in which small peasant farms prevailed was intensive agriculture, in its various forms. Examples of this are often taken from wine-producing areas, where these kinds of properties and inheritance practices can be documented.[34] Behind this argument is the notion that preindustrial agriculture was generally-speaking inelastic, and permitted property divisions only after external inputs (such as the introduction of new lucrative crops) had led to increased production. If one begins with a more realistic estimate of agricultural potential, one could argue, however, that it was precisely the subdivision of property that made such intensification processes necessary. But in this case as well an important qualification must be made: partible inheritance as an institution was not synonymous with the increasing fragmentation of farms; it was in the first place a means of distributing power within the family. Fragmentation occurred only when other conditions were also present, the most important being demographic growth. While one version is based on dubious economic hypotheses, the other requires the assistance of additional economic hypotheses. On an empirical level, the drawbacks of the intensity thesis are exposed by the fact that, in our case, neither territories characterized by small peasant farms nor those with large ones were linked to ecological zones; rather, they stretched from the southern German area to high-altitude Alpine regions (see maps 3.1, 6.1, 8.1).

Among scholars, the thesis that partible inheritance was favored by clustered settlement patterns is just as widespread as the intensity thesis, and with more valid reasons, in my view. The two circumstances could occur together, though not necessarily so. While the correlation with intensity is based on certain hypotheses regarding potential partible inheritance, the correlation with settlement patterns is conceptual and practical. In contrast to what happened in isolated hamlets, farming establishments in villages with fragmented parcels were not contiguous spatial unities whose maintenance could have constituted an argument in discussions about inheritance. With each successive hereditary transfer, houses and parcels could be recombined, without making the farms smaller or creating additional transport problems. I think that this relationship between settlement pattern and inheritance can best be generalized for the entire Alpine space if it is placed in historical perspective. After having gone through juridical solidification and territorialization, inheritance norms grew in importance along with the growth of state structures. These norms acquired greater autonomy with respect to local customs and themselves became a factor affecting the further development of settlement patterns.[35]

Such processes affected many aspects of social life, from relations of stratification to household hierarchy and the gendered division of labor, to ideas about kinship and public roles. These phenomena are clearly illustrated by a well known anthropological study, dating from the 1960s, which examines two localities on the border between the Trentino and the Alto Adige/Südtirol (within the Tyrolean transitional zone described above).

In the Val di Non, just below the pass linking the valley to the Etschtal, a settlement of German-speaking peasants was established during the late Middle Ages, a few kilometers above a village of Italian speakers that had been founded in ancient times. Throughout the existing historical memory of the area, ideas about family and kinship held within each community diverged widely. In the more recent settlement, farms were inherited undivided in a way that corresponded to the emphasis on the farming establishment, a perspective that had been supported by the Tyrolean princes. Authority was concentrated in the person of the farm owner or the eldest son. Family members who lived in the house were mostly excluded from decision-making and their status almost resembled that of servants. Kin who lived elsewhere had few interactions with the family, neither for work nor socially. Thus, the farmer-owners were strongly oriented toward communal politics and its institutions. The Trentino village offered a different picture, in which inheritance practices that treated sons and daughters equally were part of the official system of values. This was paired with diffuse patterns of authority between brothers and sisters, and the intense maintenance of contacts with kin. Personal relationships were more important that formalized ones, and in the political sphere patron-client relations were openly prevalent.[36]

State Formation and Society

Anthropological studies have enriched much of our historical understanding of families and domestic units, but often, when analyzing communes and lordship, they rely on concepts that are focused on autonomy in a one-sided way, and thus often generate unrealistic results. Such approaches are especially common in studies of the Alpine regions, where communal formations played an important role during the early modern period, such as in the *Grand Escarton* of Briançon in the western Alps, and in the rural cantons and republican communes of what is now Switzerland. An autonomistic perspective would view the commune as the product of a principle of horizontal solidarity that grew out of conflict with noble lords and other external hierarchs or even in self-defense against 'feudalism.' Communal structures were therefore evaluated according to their relative ability to "resist economic and political pressure from outside."[37]

But the formation of communes and associations of communes can not be understood independently of the general acceleration in the process of state formation during the early modern period. Only then can one observe the commune's second key function, one which associates it with other constitutional forms: its ability to impose an internal order and political structure. From this perspective, the strengthening of the commune resulted in a change in dominion. The region of the Three Leagues, which we have examined above, is a good example, organized as it was mainly on communal structures. The nobility did not cease to exist there during the sixteenth century, but it developed and presented a new face. These nobles, from then on, self-consciously turned into the elite of the emergent Freestate. Economic inequality and social control effected through the communes and their political leaders were not insignificant, even if political decision-making involved a large portion of the male population. In a memorial written early in the eighteenth century one reads that communes were very independent in this country: "The magistrate is its sovereign, and the nobility leads its communities."[38]

Some authors are convinced that the Alpine environment favored communal formations and created obstacles for feudal lords, since agricultural surpluses there were not attractive enough for the aristocracy.[39] From what can be gleaned from observable developments, this claim is neither true nor false. During the early modern period, in some parts of the central and western Alps, the communes played an important role, while in the eastern Alps, for the most part, they played only a subsidiary role behind the noble lordships. But even when the claim appears to be true, I believe it to be exceedingly static.

Chapter Eight

In the first place, agricultural production is not a simple function of geography. Economic differences between the Alpine space and surrounding areas grew larger alongside the general intensification of agriculture during the eighteenth and nineteenth centuries (see chapter 3), but they had been substantially smaller during the earlier period. In the second place, the nobility did not remain tied to agricultural output: depending on the context of the state in question, nobles could assume a markedly urban character. Their absence from the mountains was therefore only indirectly related to the issue of the agrarian surplus, but very directly related, on the other hand, to the attractiveness of urban centers (see chapter 4). French historians have explained how difficult it is, in their part of the Alps, to distinguish precisely between communal and feudal-lordly elements, and they have shown that during the seventeenth and eighteenth centuries these mixtures are not consistent from region to region. For the Dauphiné, Bernard Bonnin has reached the conclusion that "the lords, whether exercising jurisdiction or simply holding land rights, were less numerous in mountain areas than in the lowlands, except for the mountainous areas closest to Grenoble, where the urban nobility—and especially the *parlementaires*—were well established." This distribution was the result of a process in which the power of urban attraction became progressively more important. Here, during the Old Regime, the lesser nobility also fell increasingly under the spell of city life, permitting new social groups to assert themselves in the mountainous zones.[40]

This brings us to the oft-articulated, oft-discussed topic of freedom. As has been pointed out in the introductory chapter, there are good reasons to suppose that the Alpine population disposed of particular "liberties" in the old sense of the word. "Rights and liberties," regularly-used expressions in the society of orders, defined the specific status of local, regional, or other associations. Because of the fact that, from the beginning of the state-building process, power centers have been for the most part located either at the margins or outside of the Alpine space, the level of autonomy experienced in many mountain regions was greater than usual. Their proudly proclaimed l"aws and liberties" were the flip side of their distance from the seat of power (see chapter 1).

During the early modern period such liberties could be discussed without reference to equality. The concept of "liberties" was also valid for territories in which they were borne by the organized nobility, such as in Carinthia. But elsewhere as well the Alpine elite was the chief protagonist of autonomy, according to a comparative study by Pierangelo Schiera. This older use of the concept of "liberty" was specific to a certain

State Formation and Society

social order, and different from the general modern political meaning of the word, which moved to the center of social attention with the Enlightenment and imposed itself along with a society of state citizens. If the Alps or certain Alpine regions (often the Swiss ones) were then held to enjoy a particular measure of liberty, or freedom, this occurred above all because such an assessment was a useful foil for an Enlightenment critique of society and civilization.[41] This sort of symbolic appropriation on the part of intellectuals from abroad, designed for consumption abroad, relied on contrasting images, and conclusions concerning their relation to reality should take this into account.

"It is easy to see that this portrait points to the perfect equality of Alpine people, where there is no nobility and not even a governor, where the possibility of career advancement does not trouble the soul and ambition is a foreign word." Thus wrote Albrecht von Haller in his didactic poem on the Alps in 1729, which also exalted Alpine liberty —fifty years later the work was in its eleventh edition. During the prerevolutionary and revolutionary periods, the link between Alps and liberty/freedom became commonplace. Immanuel Kant meditated on the mountain-dwellers' love of freedom, Friedrich Schiller located freedom in the mountains, and Napoleon Bonaparte, according to his own statements, was especially interested in the mountainous part of Switzerland. The symbolic character of this concept became particularly evident when the process of verifying it was itself carried out through symbolic means. The French writer Louis-Sébastien Mercier observed in 1785 that, after having visited a three-dimensional model of the region around the Lake of Lucerne in the town of the same name, he then understood why despotism had never been able to spread in the mountains: the subject could simply climb a bit higher up the mountainside, and he would then be able to trample the head of his oppressor under his feet. In general, though, it was enough to connect the virtue of the mountain population to their close and special relationship with nature. In 1792, for example, a delegate to the Convention in Paris stressed how the Savoyards, being so tied to nature, were the living proof that "the man of the mountains" was truly "the man of Freedom."[42]

In the nineteenth century, the range of interpretations and representations grew wider. The Alps were then discovered by, among others, the monarchies and held up by them as the place whose inhabitants were the faithful defenders of tradition against the disorders of the age. The comments of Archduke Johann of Austria were not atypical; in 1822 he affirmed having found in the mountains "vigor, fidelity, simplicity, and an as yet uncorrupted generation." His objective was to construct, by placing together

Chapter Eight

the best people and the mountain population, a defense against the insincerity of his times and "while everything else is agitating so restlessly, to be a peaceful model of how things should be everywhere."[43]

When in the middle of the twentieth century Fernand Braudel drew attention to the freedom of the mountains, he was making a piece of the Enlightenment heritage his own. "It is only in the lowlands that one finds a close-knit, stifling society, a prebendal clergy, a haughty aristocracy, and an efficient system of justice. The hills were the refuge of liberty, democracy, and peasant 'republics'". And, he explained, the feudal seigniory—as a political, social, and economic system—could not encompass the bulk of the mountain zones, and even if it reached them its influence was incomplete. According to Braudel, this was the case wherever the population was thin and dispersed. In my view, such generalizations are unwarranted, and interesting only in the degree to which they suggest a relationship between population density and seignorial structure. The density differential between the Alps and the surrounding areas already existed at the start of the early modern period, and increased ever more rapidly over the course of our study period. But does the existence of democratic and republican political forms derive directly from population data? Is this argument not equally as defective as that used by authors who find in the low demographic density of eastern Europe the explanation for a seignorial system there, a system whose feudal character is emphasized by all?[44] If certain demographic factors are sometimes held to explain the absence of a "haughty aristocracy" and at other times its presence, it makes more sense to approach the question differently and to grant political factors the influence that is rightly theirs. By so doing, we do not compromise our ability to generalize: in the sphere of politics it is also possible to identify regular patterns.

We can summarize by saying that, beginning in the sixteenth century, the acceleration in the process of state formation led to a growth in the pre-existing structural differences between different areas. Regional developmental paths were characterized by the relative power relationships and configurations that prevailed at various social levels. As far as the level of farming establishments—an under-researched topic—was concerned, there were readily apparent similarities between Alpine structures and those in the surrounding areas. The prevalence of large peasant farms in the eastern Alpine space differentiated this region from other parts of the Alps, but linked it directly to Danubian Bavaria and Upper Austria. A broader context is necessary to understand other dimensions of development as well. The specific impact of the older "liberty" in the Alps was related to the fact that, from the beginning of the state-building process, the power centers were not located in mountain regions, but in nearby or far-off lowlands. Enlightenment discourses about freedom, carried forward by intellectuals who lived in such centers, turned the greater autonomy of the mountains into a critical mirror for their own societies. In the modern bourgeois sense, the generic "freedom" of the Alpine population was, however, a literary construction that was disconnected from concrete historical context.

Even today traces of this multiform sociopolitical history can be found. Consider, for example, the diversity of the countries that signed the Alpine Convention in 1991: two republics with centralizing traditions (France, Italy), three republics with a more or less federalist traditions (Switzerland, Germany, and Austria), one republic with a Socialist past (Slovenia), and two dynastic principalities (Liechtenstein, Monaco). The practical foundations according to which these states carry out their agrarian policies and their support of Alpine agriculture once more mirror an older bipartition of the Alpine space. Countries in the eastern Alps (Austria, Slovenia) rely on a cadastre of individual farms, while in all of the other regions the mountain area is territorially defined on a communal basis.[45]

Endnotes

1   Paul and Germaine Veyret, *Au cœur de l'Europe - les Alpes* (Paris, 1967), 263.
2   Fernand Braudel, *Civilisation matérielle, économie et capitalisme, XVe–XVIIIe siècle* (Paris, 1979), esp. 2: 231, 244.
3   Boserup 1993, 77–87.
4   *Geschichte der österreichischen Land- und Forstwirtschaft und ihrer Industrien 1848–1898* (Vienna, 1899), 1: 49; Dieter Schwab, "Eigentum," in *Geschichtliche Grundbegriffe. Historisches Lexikon zur politisch-sozialen Sprache in Deutschland* (Stuttgart, 1975), 2: 70–72, 89–92.
5   G. Chittolini and G. Coppola, "Grand domaine et petites exploitations: quelques observations sur la version italienne de ce modèle (XIIIe–XVIIIe siècles)," in Gunst and Hoffmann 1982, 175–92.
6   See chapters 2–4 above.
7   Kain and Baigent 1992, 181–90, citation from 187; Renato Zangheri, "I catasti," in *Storia d'Italia*, ed. Giulio Einaudi (Turin, 1973), 5: 759–806; Carlo Capra, "The Eighteenth Century. The Finances of the Austrian Monarchy and the Italian States," in *Economic Systems and State Finance. The Origins of the Modern State in Europe*, ed. Richard Bonney (Oxford, 1995), 295–314; for Savoy see chapter 7 above.
8   Ruggiero Romano, "La Storia economica. Dal secolo XIV al Settecento," in *Storia d'Italia*, ed. Giulio Einaudi (Turin, 1974), 2: 1811–1931; Domenico Sella, *Italy in the Seventeenth Century* (London and New York, 1997), 34, 63–69.

Chapter Eight

9   Giorgio Chittolini, "Feudalherren und ländliche Gesellschaften in Nord- und Mittelitalien (15.–17. Jahrhundert)," in *Klientelsysteme im Europa der Frühen Neuzeit*, ed. Antoni Maczak (Munich, 1988), 243–59; Chittolini 1988; Chittolini and Coppola 1982 (see n. 5), 184–85; regional examples in Scaramellini 1978, 54–58, 164; Hugo Penz, *Das Trentino. Entwicklung und räumliche Differenzierung der Bevölkerung und Wirtschaft Welschtirols* (Innsbruck, 1984), 170–74; depending on the specific starting points and political situations, nineteenth-century agrarian reforms in northern Italy were somewhat less drastic, see *Storia dell'agricoltura italiana in età contemporanea*, ed. Piero Bevilacqua (Venice, 1990), 2: 45–103.

10 Adam Christian Gaspari, *Vollständiges Handbuch der neuesten Erdbeschreibung. Mit genauer Bemerkung der wichtigsten Ereignisse und politischen Umwandlungen* (Augsburg, 1802), 1: 550; Heide Wunder, "Das Selbstverständliche denken. Ein Vorschlag zur vergleichenden Analyse ländlicher Gesellschaften in der Frühen Neuzeit, ausgehend vom 'Modell ostelbischer Gutsherrschaft'," in Peters 1995, 23–49, esp. 25–27.

11 Wunder 1995 (see n. 10), 24.

12 For the discussion of Germany and eastern Europe, see for example Peters 1995; Holenstein 1996; Heide Wunder, "Agriculture and Agrarian Society," in *Germany. A New Social and Economic History*, ed. Sheilagh Ogilvie and Bob Scribner (London, 1996), 2: 63–99; *The Origins of Backwardness in Eastern Europe. Economics and Politics from the Middle Ages until the Early Twentieth Cenury*, ed. Daniel Chirot (Berkeley, 1989).

13 Tilly 1992, esp. 1–65, 127–60 187–88, citation from 9; for a brief synthesis see id., "The Long Run of European State Formation," in Blockmans and Genet 1993, 137–50.

14 Tilly repeatedly draws attention to the significant differences between agricultural regions (Tilly 1992, 23, 27, 33–34, 48, 152), but his models fail to distinguish between agriculture and coercion; if the two were congruent, then the entire, scarcely urbanized Alpine space (not to mention other regions) would have to be defined as "coercion-intensive"—which would be ironic, given the clichés about Alpine liberty. An important influence for his model was Barrington Moore, *Social Origins of Dictatorship and Democracy. Lord and Peasant in the Making of the Modern World* (Boston, 1966).

15 Tilly 1992, esp. 4, 25–26, 29, 58, 137, 160–87.

16 Herbert Knittler, "Zwischen Ost und West. Niederösterreichs adelige Grundherrschaft 1550–1750," in *Österreichische Zeitschrift für Geschichtswissenschaften* 4 (1993): 191–17, citation from 193; since the lord was also a public authority, the modern term employed by Knittler and others to refer to various feudal dues and contributions ("rent") is inadequate, because it fails to capture the"tax "aspect of the subjects' obligations and invites economistic interpretations of power relations; for labor obligations (*corvées*) see Renate Blickle, "Scharwerk in Bayern. Fronarbeit und Untertänigkeit in der Frühen Neuzeit," in *Geschichte und Gesellschaft* 17 (1991): 407–33.

17 Knittler 1993 (see n. 16), 203, 205–6, 210–11; the role of the (international) market in the development of direct management by the lord is among the especially controversial topics, see Holm Sundhaussen, "Zur Wechselbeziehung zwischen frühneuzeitlichem Aussenhandel und ökonomischer Rückständigkeit in Osteuropa: Eine Auseinandersetzung mit der 'Kolonialthese'," in *Geschichte und Gesellschaft* 9 (1983): 544–63.

18 Knittler thus qualifies some of the other suggestions for a three-way (rather than two-way) division of the area; see Friedrich Lütge, *Geschichte der deutschen Agrarverfassung vom frühen Mittelalter bis zum 19. Jahrhundert* (Stuttgart, 1967), 171.

State Formation and Society

19   Eastern lordly estates, for example, were based not on the obligatory service of "small farms" (as in Tilly 1992, 152); rather, farm sizes ranged from roughly 30 to 70 hectares—see H. Harnisch and G. Heitz, "Feudale Gutswirtschaft und Bauernwirtschaft in den deutschen Territorien. Eine vergleichende Analyse unter besonderer Berücksichtigung der Marktproduktion," in Gunst and Hoffmann 1982, 24.

20   F. Irsigler, "Gross- und Kleinbesitz im westlichen Deutschland vom 13. bis 18. Jahrhundert: Versuch einer Typologie," in Gunst and Hoffmann 1982, 33–59; for the statistics see Barthel Huppertz, *Räume und Schichten bäuerlicher Kulturformen in Deutschland. Ein Beitrag zur Deutschen Bauerngeschichte* (Bonn, 1939).

21   David Warren Sabean, *Property, Production, and Family in Neckarhausen, 1700–1870* (Cambridge, 1990), esp. 13–17, 26–27, 43–44, 71–73, 185–87, 248–49; also id., "Aspects of Kinship Behaviour and Property in Rural Western Europe Before 1800," in *Family and Inheritance. Rural Society in Western Europe 1200–1800*, ed. Jack Goody et al. (Cambridge, 1976), 96–111.

22   An important source for assessing direct lordly management are the so-called *Robot* decrees of the late eighteenth century; see Knittler 1993 (see n. 16), esp. 196–99; for the eastern rim of the Alps, see also Helfried Valentinitsch, "Gutsherrschaftliche Bestrebungen in Österreich in der frühen Neuzeit. Unter besonderer Berücksichtigung der innerösterreichischen Länder," in Peters 1995, 279–97. Evidence concerning forms of inheritance deal with transmission of land, and are based on investigations carried out at the end of the nineteenth century and at the beginning of the twentieth; see Ingrid Kretschmer and Josef Piegler, "Bäuerliches Erbrecht. Kommentar," in *Österreichischer Volkskundeatlas*, fasc. 2 (Graz, 1965), 1–18. On map 8.1 various mixed regions and western areas with basically impartible inheritance regimes are not indicated; in the Alps the line follows the small map of Kretschmer and Piegler (in the work cited above), and in southern Germany it follows map 1 in Huppertz 1939 (see n. 20). Some Alpine regions, on the northern slope of Switzerland and in southern France, are more or less exceptions to the predominant rule of real partibility in the central and western Alpine space; see Arnold Niederer, Bäuerliches Erbrecht. Kommentar, in *Atlas der schweizerischen Vorkskunde*, pt. 1, fasc. 7 (Basel, 1968), 570–600 (also for the Swiss *Mittelland*) and Dionigi Albera, "Familles. Destins. Destinations. Entre mosaïque et portrait-robot," *Le monde alpin et rhodanien* (1994): 7–26 (critique of the traditional classification).

23   A graphic representation of the network of alliances and conflicts is found in Tilly 1992, 176; also *Monarchs, Ministers and Maps. The Emergence of Cartography as a Tool of Government in Early Modern Europe*, ed. David Buisseret (Chicago and London, 1992), 101–102.

24   See chapters 6 and 7 above.

25   Wilhelm Abel, *Geschichte der deutschen Landwirtschaft vom frühen Mittelalter bis zum 19. Jahrhundert* (Stuttgart, 1967), 69; Lutz K. Berkner and Franklin F. Mendels, "Inheritance Systems, Family Structure, and Demographic Patterns in Western Europe, 1700–1900," in *Historical Studies of Changing Fertility*, ed. Charles Tilly (Princeton, 1978), 209–23, here 212.

26   Salzburg provides a good example of farm-centered development under princely influence; see *Geschichte Salzburgs* 1983–1991, 1: 629–33 and 2: 2506–29, 2559–61; essentialist hypotheses, whether regarding peasant and lordly preferences for partibility, or the productivity of small or large farms, weaken many historical studies.

27   See chapter 6 above; in order to prevent misunderstandings, it should be clarified that "region" and "transitional region" are constituted here on a comparative basis; from a different observational perspective, another set of boundaries would likely be obtained.

Chapter Eight

28   *Geschichte der österreichischen Land- und Forstwirtschaft* 1899 (see n. 4), 1: 285–309; Stolz 1949, 323–27, 435–49; Hermann Wopfner, *Bergbauernbuch. Von Arbeit und Leben des Tiroler Bergbauern in Vergangenheit und Gegenwart*, fasc. 2 (Innsbruck, 1954), 133–76; often the scholarship, retroactively, takes up a position in defense of impartible inheritance; for the cadastre see also Kain and Baigent 1992, 192.

29   Scholarship can be said to have moved beyond ethnic theories concerning settlement patterns. These theories grew and spread widely under the influence of nationalism, and they unabashedly projected their notions of "peoples" into the past; see for example Rainer Loose, "Siedlungsgeschichte des südlichen mittleren Alpenraumes (Südtirol, Trentino, Bellunese) seit der Karolingerzeit. Ein Überblick," *Tiroler Heimat* 60 (1996): 5–86. Obviously this does not mean that environmental and cultural factors did not exercise influence within a shorter-range, historically specific context.

30   Bernard Derouet, "Pratiques successorales et rapport à la terre: les sociétés paysannes d'Ancien Régime," *Annales ESC* (1989): 173–206, citation from 185; with respect to the problematic treatment of Alpine societies and their forms of inheritance, see Albera 1995, esp. chapter 4.

31   See chapter 3 above.

32   Mitterauer 1986, 185–324, citation from 192; Mitterauer 1990, 131–45, citation from 137; regarding the set of scholarly issues tied to his eco-typological approach, see also chapter 5 above.

33   See Jon Mathieu, "Von der verstreuten Familie zum 'Ganzen Haus'. Sozialgeschichtliche Übergänge im schweizerisch-österreichischen Alpenraum des 17. bis 19. Jahrhunderts," in *Der Vinschgau und seine Nachbarräume*, ed. Rainer Loose (Bolzano/Bozen, 1993), 245–55, here 247–48.

34   For example, Mitterauer 1990, 144; in a very specific sense, see also Sabean 1990 (see n. 21), 15–16.

35   For a bibliography and materials related to settlement issues, see Mathieu 1992, 66–73; it is not our intention here to discuss the medieval genesis of settlement forms; for the modern distribution of settlements, see Günter Glauert, *Die Alpen, eine Einführung in die Landeskunde* (Kiel, 1975), 72, 81.

36   Cole and Wolf 1974, esp. 233–62; this study also stresses the difference between these ideologies and practice; for this issue see the important observations of Albera 1995, chapter 14.

37   Thus writes Viazzo 1989, 296; examples of similar perspectives are in Rosenberg 1988 and Albera 1995, chapter 16; for the historical discussion see Holenstein 1996, 75–81.

38   See *Regesten und Register zu den Acta Helvetica*, ed. Kurt-Werner Meier et al. (Aarau, 1996), no. 27 and also chapter 7 above. It is difficult to assess the level of economic inequality in a given territory on the basis of a few outlying communes, see for example the study of Randall McGuire and Robert McC. Netting, "Levelling peasants? The maintenance of equality in a Swiss Alpine community," *American Ethnologist* 9 (1982): 269–90.

39   One example is Perry Anderson, see the introduction to chapter 6 above.

40   Pierre Léon et al., "Régime seigneurial et régime féodal dans la France du sud-est: déclin ou permanence? (XVIIe–XVIIIe siècles)," in *L'abolition de la féodalité dans le monde occidental* (Paris, 1971), 1: 147–68, and for a discussion ibid., 2: 609–26; Bernard Bonnin, "L'élevage dans les hautes terres dauphinoises aux XVIIe et XVIIIe siècles," in *L'élevage* 1984, 263–81, citation from 269.

41   Werner Conze et al.,u.a. "Freiheit," in *Geschichtliche Grundbegriffe. Historisches Lexikon zur politisch-sozialen Sprache in Deutschland* (Stuttgart, 1975), 2: 425–542; Pierangelo Schiera, "L'autonomia locale nell'area alpina. La prospettiva storica," in Schiera 1988, 3–50, 149–54, here 152; Richard

Weiss, *Die Entdeckung der Alpen. Eine Sammlung schweizerischer und deutscher Alpenliteratur bis zum Jahr 1800* (Frauenfeld, 1934).

42   Albrecht von Haller, *Die Alpen und andere Gedichte* (Stuttgart, 1965), footnote at verse 100; Andreas Bürgi, "Höhenflüge," in *Die Schwerkraft der Berge 1774–1997*, ed. Stephan Kunz et al. (Basel, 1997), 33–35; Nicolas 1989, 168.

43   Othmar Pickl, "Wirtschaft und Gesellschaft in den Ostalpenländern Kärnten und Steiermark bis zur Mitte des 19. Jahrhunderts," in Mattmüller 1986, 38–101, citation from 88–90.

44   Braudel 1972, 1: 38–41 ("Mountain freedom"), citation on 1: 40; Holm Sundhaussen, "Die Ursprünge der osteuropäischen Produktionsweise in der Frühen Neuzeit," in *Die Frühe Neuzeit in der Geschichtswissenschaft. Forschungstendenzen und Forschungserträge*, ed. Nada Boskovska Leimgruber (Paderborn, 1997), 145–162.

45   Bätzing 1991, 102, 113.

9 History of the Alps from 1500 to 1900

A summary

"As a European heritage the Alps form a natural, historical, cultural, and social entity of vital significance," states an action plan for the "Future of the Alps" that was adopted by an international symposium in 1974. Does this mean we should consider the Alps—contrary to historiographic tradition—not merely as a geographic entity but as a historical one as well? In one form or another, this question has been at the center of many recent debates about possible new approaches. In attempting to answer it we have to consider the actual development of the mountain area and to include the surrounding lowlands as a basis for comparison. Here, we examined a period covering four centuries, from the end of the Middle Ages to 1900, from an economic and a sociopolitical perspective.

The first part of the study (chapters 2–5) deals with the *relations between population growth, economic development, and the Alpine environment.* The Alpine population nearly tripled during the period under consideration. Within a certain perimeter, it grew from an estimated 2.9 million in 1500 to roughly 7.9 million in 1900. Demographic growth and high percentages of agrarian workers suggest that agricultural production increased substantially during the period. The intensification of agriculture was the result of an increase in cropping frequency and of changes in livestock and crop selection (more cattle, introduction of corn and potatoes). In general these new, land-saving ways of using environmental resources augmented the workload considerably; as a consequence, many intensification options only came into use as population pressure increased consumption levels and labor capacity.

To get an idea of the progress of urbanization, consider the emergence of larger cities. Around the year 1500 the area under investigation possibly included just one city with a population exceeding 5000 inhabitants. Around 1800 that number had grown probably to 9 and in 1900 to 42. Demographic-agrarian and urban developments cannot be understood separately. The potential for urban growth was influenced by population density and agricultural intensity, and urban growth, in return, affected the population and the agriculture of the surrounding area.

But what about the fact that the Alpine environment faced undoubted limits regarding its potential for exploitation? Agronomists estimate that annual grass yields decrease by 40 percent per 1000 meters of altitude. Yet the effects of this limited potential varied over time. In a first phase (sixteenth and seventeenth centuries) the growth difference, both between high and low Alpine regions and between the Alps and adjacent flatlands, was much smaller than in a second phase (eighteenth and nineteenth centuries). While in the first phase some mountain regions had higher growth rates than the adjacent flatlands, the second phase was marked by a generally faster and accelerated flatland growth. The impact of altitude on development increased with agricultural intensity—it was, in other words, the result of a historical process. By 1900, the Alpine area, in its economic dimensions, differed much more clearly from its surroundings than at the beginning of the modern period.

The second part of the study (chapters 6–8) examines the *impact of political factors on rural conditions and society*. As regional examples from the western, central, and eastern Alps show, the formation of state and society in early modern times diverged considerably in different regions. In Savoy and in the Grisons the increasing power of princes and communes, respectively, favored centralist and localist structures, whereas in Carinthia the intermediate social force of the nobility was strengthened. On the level of peasant households the differences were no less distinct. In the first two regions small holdings with male or gender-neutral forms of partible inheritance prevailed while big farms and impartible inheritance were typical in the east.

Since the late eighteenth century the large-scale forces of the state increased their influence on regional and local power structures. Despite marked regional differences

Chapter Nine

in chronology and relevance, the abolition of feudal rights led to a modernization of property relations almost everywhere. Still, even around 1900 the agrarian structure of the Alps was far from uniform. According to statistics—by then available on a large scale—smallholders were the rule in the west and center of the Alpine arc while the east continued to have a large percentage of big farms with numerous farmhands. These differences between east and west were not limited to the mountains but extended to the adjacent lowlands as well. The big farm structure of the eastern Alps, for instance, extended north into the Bavarian and Upper Austrian regions of the Danube area. Political interpretations of agrarian structures, thus, gain in plausibility relative to environmental or economic approaches.

Generally speaking, territorial state institutions consolidated during the sixteenth century and brought about an increase in structural differences. Of crucial importance for the different regional development paths was the balance of power between princes, lords, and communes, and their respective forms of organization around 1500. Wherever one of the forms clearly prevailed at that time, chances were it would extend its influence and dominate the other forces in the process of state building. A similar differentiation, although less marked, occurred on the level of household and farming. For conceptual and practical reasons, village-type settlements tended toward partible inheritance and potential small holdings, whereas isolated farms had a larger individual expansion potential and encouraged integral property transfers. Each of these configurations favored a different development when legal norms were promoted and extended by state building. Yet the territorial norms now increased the autonomy of the law from local custom and allowed it to become a factor in the evolution of settlements.

The question, then, as to whether or not the Alps constitute a historical entity, can be answered in more than one way. In some vital economic aspects, from the end of the Middle Ages until 1900, the Alpine trajectory progressively diverged from that of the surrounding flatlands. Alpine cities—few in numbers and growing more slowly than in non-mountain areas—are a telling example. In some crucial sociopolitical aspects, on the other hand, certain Alpine regions bore more resemblance to adjacent lowland territories than to other Alpine regions. The East-West differences in the agrarian structure are a case in point. A third answer is to be found on the methodological level: a historical entity is, after all, not just an area with a "common history" but quite simply an area studied by historians. Having chosen the Alps as the object of their research, they encounter an area with a high density of borders and a corresponding number of different historiographic traditions. The comparative study of such nationally-colored traditions is a rewarding task that may contribute significantly to the general culture of the field.

Arguments and outlook

At various points this study has been concerned with theoretical issues. Here I will mention a few of the assumptions and discursive approaches whose prominent place in the literature needs to be revised, in my view.

Adaptation to the environment (chapter 5). As has been mentioned, the problem with the notion of adaptation is its very adaptability in changing historical circumstances. Small populations exploit mountain lands differently than do large groups, which perhaps produce for long-range markets. One can assume that at each step of the process of agrarian intensification, preference was shown for those parcels of land that seemed particularly well suited to given conditions. While examining the advantages and disadvantages of a specific use of resources had always been part of peasant practice, it is impossible to deduce, in a fixed way, from terrain conditions and the environment, how the land was used at any given point in time. In this sense, it is misleading to speak of adaptation to particular natural conditions. Historical relativity also applies to livestock-raising and Alpiculture, which are often represented as the perfect example of an adaptive use of the mountain environment. But this overlooks the fact that livestock-raising, in contrast to crop cultivation, is carried on over the course of an entire year. Animals had to be wintered in some way, and this is not easily reconciled with the idea of a natural economic arrangement.

Modest resources and tight growth limits (chapter 5). Assessments of environmental potential are influenced by the way in which the object of study is defined. Many Alpine studies focus on the demography of small localities during the brief chronological periods for which good documentation survives. It is not unusual that these studies undergo a slippage by which the methodological boundaries of these study areas become factual boundaries, and the short-term demographic and economic situations become the absolute measure of the territories' productive capacity. In this way, resource scarcity and its impact on demographic phenomena become an important theme, while the intensification processes sparked by population growth go largely unnoticed. If one chooses to frame these developments differently, one sees a significantly modified picture. The productive potential of the entire Alpine space between the sixteenth and nineteenth centuries should not be underestimated. For the first two centuries, it appears that growth in this area hardly differed at all from that of the flatlands, and while this growth continued during the eighteenth and nineteenth centuries, the differential between Alpine and lowland rates began to increase. Compared to nearby areas, what came to be missing in the mountain regions was above all a temporal resource: the altitude-specific length of the growing season became a critical factor as agrarian intensification continued to progress.

Population density, economics, and politics (chapters 5 and 8). More attention needs to be given by historical studies to population density and its effects. Already during the Middle Ages the Alpine realm was generally less densely inhabited and urbanized than nearby regions, and as time went on this inequality grew significantly. Many economic relations between the Alps and surrounding areas can be examined from the perspective of this density differential (wood transport, commerce in livestock, migration). Consideration of density is also useful for understanding certain sociopolitical phenomena. For example, the centers of the major state formations were not located within the Alpine space, but were almost all in neighboring regions, at varied distances. Their spatial remove from such power centers, accompanied by their noteworthy local and regional autonomy, was thus characteristic of a whole series of mountain regions. But low levels of population density and "liberties" in the old sense of the term do not permit one to draw automatic conclusions about internal power structures, which could have been determined either by an organized nobility or by communities and their elites.

Agrarian intensity and inheritance patterns (chapter 8). Agricultural intensification is often included among the explanations for the emergence of property division and small peasant farms. Relevant examples are taken, for instance, from grape-growing zones in which such proprietary and inheritance forms are historically documented. This kind of argument is based on the notion that preindustrial agriculture was generally inelastic and permitted property divisions only following a prior increase in production due to external factors. However, one would be more justified in arguing that it was precisely the division of holdings that made these intensification processes necessary. Still, both explanations lack precision in that property divisions were not synonymous with the fragmentation of peasant farms. Division was in the first instance a way of distributing power within a family. Fragmentation occurred only when other conditions—population growth above all—held as well. It is thus not surprising that the territorial diffusion of various forms of agriculture and inheritance differed widely across the Alpine space.

Communities and lordship (chapter 8). Among the scholarship on communities and lordship, there is some work that takes as its starting point concepts about autonomy, and often generates one-sided results. This is above all the case for mountain regions, where communal associations during the early modern period played an important role such as in certain regions of the western and central Alps. From an autonomistic perspective, the commune appears to be the manifestation of a principle of 'horizontality' that asserted itself during the struggle with outside lords and hierarchies, in self-defense against 'feudalism.' Communal structures are measured by their ability to resist external political and economic pressure. But the formation of communes and communal associations can not be understood independently of the state building process. This brings to light the ability of the communes to impose an internal order and power structure that was not dissimilar from the constitutional configuration of other political formations. Seen from this perspective, the reinforcement of the commune was one particular variation of a shift in lordship. The nobility modified its frame of reference as this shift occurred, but its historical existence was not at risk.

Taken together, our findings make it possible to reflect on the relationship between economic and sociopolitical development. Alpine history from the sixteenth through the nineteenth centuries offers examples of how economic and sociopolitical processes were at the same time dependent and independent. This quasi-open relationship can be illustrated with a model that takes as a key variable the issue of population density, as discussed above. The number of people present in a given territory at a certain moment during the study period tells us a great deal about their economic requirements and potential, and something about their organizational needs and possibilities as well. But the distribution of the population across the territory is not determined by density. Doesn't it make a significant difference, precisely with respect to sociopolitical phenomena, whether people gathered in centers or spread themselves out over the entire area?

Years ago, in the museum of a small community on the Dutch Frisian coast, I learned a lesson about the fascination with the Alps that was as unexpected as it was unforgettable. Surrounded by a landscape that had been seized from the sea and that could not be any flatter, an eighteenth-century bourgeois family had had its salon painted on all four walls, from the ceiling to the floor, with an Alpine scene that could be identified with precision. The pretend Alpine *Stube* from Hindeloopen is but one among many examples of the power of imagination, which has both inspired and hindered

Chapter Nine

research on the Alps. In 1761 a novel was published in Amsterdam which enjoyed an unheard-of success and was titled *Julie, ou la nouvelle Heloïse. Lettres de deux amans, habitans d'une petite ville au pied des Alpes*, by Jean-Jacques Rousseau. Together with his previous writings and especially with his later ones, it propagated an image of the Alps that appeared in contrast to a critical view of civilization. The Alps were the opposite pole, a realm of otherness, the incarnation of nature. Even though these intellectual projections have always been subject to criticism, they have rooted themselves deeply in European tradition. With the popular spread of Alpine literature and enthusiasm for the Alps during the nineteenth and twentieth centuries, the Rousseauian image of contrast embedded itself in the European cultural patrimony on a widespread scale. It refracts itself multiple ways, varying from group to group and according to different points of view.

What we need today is not a new representation of contrast, but a normalization of scholarly discourse. Our overview of Alpine history shows us that the Alps can be differentiated from their surrounding lowlands in a variety of ways that change over time, but that they have never been an other world with respect to them or to other European centers. An important basis for this illusion was the fact that most intellectuals, along with their publics, approached the Alps and its population from the outside, to satisfy their own interests. This was a case of one of those rather strange declarations of love, in which the opinion of the beloved was not an issue. The time for such approaches should be over by now. The differential in Alpine land use, and therewith what we mean by civilization and nature, has changed radically in the recent past. At the same time Alpine voices and self-interpretations demand a hearing in ways that cannot be ignored. One can thus hope that a balanced—and perhaps even democratic—forum for discussion might emerge, in which all interested parties have a place and historical reality carries as much weight as historical imagination.

Appendices

Appendices

Table A.1: Percentages of agricultural workers in Alpine regions, 1870 and 1900

Region	1868/1872		1900/1901		Change 1870–1900	
	AW	MW	AW	MW	AW	MW
1 Alpes-Maritimes (F)	45	45	30	31	−15	−14
2 Alpes-de-Haute-Provence (F)	80	81	62	66	−18	−15
3 Hautes-Alpes (F)	79	78	63	60	−16	−18
4 Savoie (F)	76	75	66	63	−10	−12
5 Haute-Savoie (F)	72	73	65	65	−7	−8
6 Imperia (I)	76	72	64	59	−12	−13
7 Val d'Aosta (I)	86	83	81	76	−5	−7
8 Sondrio (I)	87	84	82	78	−5	−6
9 Trent (I/A)	69	–	73	66	+4	–
10 Bolzano/Bozen (I/A)	68	–	70	64	+2	–
11 Belluno (I)	77	70	66	55	−11	−15
12 Valais (CH)	82	79	67	69	−15	−10
13 Ticino (CH)	60	46	46	34	−14	−12
14 Grisons (CH)	67	64	52	52	−15	−12
15 Uri (CH)	62	66	45	52	−17	−14
16 Unterwalden (CH)	56	64	46	58	−10	−6
17 Schwyz (CH)	50	58	36	50	−14	−8
18 Glarus (CH)	17	25	18	28	+1	+3
19 Appenzell (CH)	20	32	20	31	0	−1
20 St. Gall (CH)	37	44	24	33	−13	−11
21 Liechtenstein (FL)	–	66	–	52	–	−14
22 Vorarlberg (A)	53	–	42	37	−11	–
23 Tyrol (A)	68	–	62	55	−6	–
24 Salzburg (A)	63	–	57	49	−6	–
25 Carinthia (A)	75	–	69	59	−6	–
26 Styria (A)	73	–	61	54	−12	–

AW = agricultural workers as a percentage of the total working population.
MW = male agricultural workers as a percentage of all working males.

Cross-national categorization of sectors according to Paul Bairoch, *La population active et sa structure. Statistiques Internationales Rétrospectives*, vol. 1 (Brussels, 1968) (the agricultural sector includes forestry, hunting and fishing; the total working population does not include those living on rental incomes and such). Due to varied data collection methods, the percentages are not precisely comparable, however.

The regions are the same as in table 2.2, with approximate adaptation to the administrative situation in 1990: (2) 1870/1900 called "Basses-Alpes"; (6) 1870/1900 called "Porto Maurizio"; (9) 1870/1900 part of the Tyrol, comprised of the districts of Borgo, Cavalese, Cles, Primiero, Riva, Rovereto (city and environs), Tione, Trent (city and environs); (10) 1870/1900 part of the Tyrol, comprised of the districts of Bozen, Brixen, Brunecken, Meran; (11) without the Tyrolean district of Ampezzo from 1870/1900; (23) with the following districts from 1870/1900: Imst, Innsbruck (city and environs), Kitzbühel, Kufstein, Landeck, Lienz, Reutte, Schwaz; (26) with the following districts from 1870/1900: Bruck, Feldbach, Gratz (city and environs), Hartberg, Judenburg, Deutsch-Landsberg, Leibnitz, Leoben, Lietzen, Murau, Radkerburg, Weiz.

Sources: F-Statistics 1872b, tab. 8 and 1901b, vol. 2; I-Statistics 1871, vol. 3 and 1901, vol. 3; CH-Statistics 1870, vol. 3 and 1900, vol. 3; FL-Statistics 1962 (1868, *estimate* for 1901); A-Statistics 1869, issue 2 and 1900b, vol. 46.

Appendices

Table A.2: Farm sizes in Alpine regions, 1900

Region	\multicolumn{2}{	}{Farms, in absolute numbers and in *percentages*}													
	0–0.5 ha		0.5–1 ha		1–5 ha		5–10 ha		10–20 ha		20–50 ha		50+ ha		Total
1 Alpes-Maritimes (F)	19000	52	4887	17	11925	32	3403	9	1712	5	666	2	169	0	36875
2 Alpes-de-Haute-Provence (F)	9650	26	6228	13	17201	47	1810	5	3461	9	3657	10	707	2	36486
3 Hautes-Alpes (F)	7676	30	4380	18	8005	32	4871	19	2857	11	1543	6	257	1	25209
4 Savoie (F)	27552	44	11503	18	23224	37	7818	13	2217	4	809	1	325	1	61945
5 Haute-Savoie (F)	23247	40	12729	18	21462	37	8570	15	2967	5	1302	2	243	0	57791
6 Imperia (I), 1930	7860	28			12514	44	2018	7	713	3	187	1	63	0	28242
7 Val d'Aosta (I), 1930	8384	18			25576	54	4344	9	1479	3	672	1	648	1	47331
8 Sondrio (I), 1930	5092	21			12220	51	1411	6	362	2	130	1	300	1	23895
9 Trent (I), 1930	16963	27			29414	46	4028	6	1114	2	402	1	526	1	63950
(A), 1902	21499	31			29620	42	4418	6	1212	2	370	1	542	1	70390
10 Bolzano/Bozen (I), 1930	3360	14	1944	8	7713	31	3700	15	3348	14	3216	13	1431	6	24712
(A), 1902	2695	11	2560	10	8106	32	3781	15	3463	14	3181	13	1479	6	25265
11 Belluno (I), 1930	8912	24	5099	14	16962	46	3718	10	1418	4	430	1	234	1	36773
12 Valais (CH)	–		2489	14	11646	67	2252	13	545	3	169	1	336	2	17437
13 Ticino (CH)	–		3579	23	10412	66	1308	8	229	1	81	1	98	1	15707
14 Grisons (CH)	–		1386	11	6659	52	2524	20	1061	8	453	4	722	6	12805
15 Uri (CH)	–		115	6	1058	58	376	21	140	8	40	2	95	5	1824
16 Unterwalden (CH)	–		157	5	1432	49	741	26	344	12	130	4	92	3	2896
17 Schwyz (CH)	–		243	6	1895	44	1272	30	545	13	204	5	110	3	4269
18 Glarus (CH)	–		194	10	851	46	438	24	215	12	46	2	112	6	1856
19 Appenzell (CH)	–		222	4	2846	56	1396	28	436	9	130	3	14	0	5044
20 St. Gall (CH)	–		1433	8	8540	50	4533	27	1994	12	340	2	187	1	17027
21 Liechtenstein (FL)	–		–		–		–		–		–		–		–
22 Vorarlberg (A)	2960	16	1683	9	7685	43	3093	17	1507	8	577	3	529	3	18034
23 Tyrol (A)	1989	6	1865	6	11634	38	5970	19	4641	15	3200	10	1652	5	30951
24 Salzburg (A)	1636	10	751	5	3429	22	2292	14	3412	22	2938	19	1367	9	15825
25 Carinthia (A)	1914	6	2184	7	8825	27	4603	14	6288	19	6654	20	2826	8	33294
26 Styria (A)	6030	7	5394	7	28846	36	14261	18	13144	16	9901	12	3638	4	81214

F: 1892, first category 0–1 ha; I: 1930, Trent and Bolzano also 1902; CH: 1905, 0–0.5 ha data not collected; A: 1902. Explanations and sources following table A.3.

Appendices

Table A.3: Agricultural labor force in Alpine regions, 1900

Region	Agricultural labor force, in absolute numbers and *per farm*						
	Owners	Family	Administrators, etc.	Agric. servants	Day laborers	Total	Farms
1 Alpes-Maritimes (F)	36799 1.0	–	191 0.0	4115 0.1	11161 0.3	–	36875
2 Alpes-de-Haute-Provence (F)	29891 0.8	–	132 0.0	5278 0.1	2767 0.1	–	36486
3 Hautes-Alpes (F)	23023 0.9	–	29 0.0	4840 0.2	1170 0.0	–	25209
4 Savoie (F)	48746 0.8	–	92 0.0	8194 0.1	2967 0.0	–	61945
5 Haute-Savoie (F)	45038 0.8	–	58 0.0	9263 0.2	4452 0.1	–	57791
6 Imperia (I), 1901	–	–	–	*4539 0.2	*11906 0.4	–	**28000
7 Val d'Aosta (I), 1901	–	–	–	*3291 0.1	*3911 0.1	–	**47000
8 Sondrio (I), 1901	–	–	–	*2812 0.1	*3292 0.1	–	**23500
9 Trent (I/A), 1902	82650 1.2	99550 1.4	1091 0.0	5273 0.1	1920 0.0	2.7	70390
10 Bolzano/Bozen (I/A), 1902	25218 1.0	37451 1.5	930 0.0	19743 0.8	2671 0.1	3.4	25265
11 Belluno (I), 1901	–	–	–	*2198 0.1	*14746 0.4	–	**36500
12 Valais (CH)	15950 0.9	33536 1.9	885 0.1	2591 0.1	2840 0.2	3.2	17437
13 Ticino (CH)	14841 0.9	28244 1.8	301 0.0	1586 0.1	871 0.1	2.9	15707
14 Grisons (CH)	11211 0.9	20022 1.6	628 0.0	4983 0.4	3175 0.2	3.1	12805
15 Uri (CH)	1919 1.1	3101 1.7	74 0.0	753 0.4	165 0.1	3.3	1824
16 Unterwalden (CH)	2889 1.0	4339 1.5	95 0.0	1126 0.4	519 0.2	3.1	2896
17 Schwyz (CH)	4113 1.0	5934 1.4	234 0.1	1305 0.3	631 0.1	2.9	4269
18 Glarus (CH)	1537 0.8	2113 1.1	14 0.0	535 0.3	185 0.1	2.4	1856
19 Appenzell (CH)	3870 0.8	2896 0.6	75 0.0	804 0.2	630 0.1	1.6	5044
20 St. Gall (CH)	14123 0.8	20523 1.2	312 0.0	4180 0.2	2111 0.1	2.4	17027
21 Liechtenstein (FL)	–	–	–	–	–	–	–
22 Vorarlberg (A)	17691 1.0	19883 1.1	805 0.0	2729 0.2	506 0.0	2.3	18034
23 Tyrol (A)	31566 1.0	52202 1.7	1465 0.0	17984 0.6	1008 0.0	3.4	30951
24 Salzburg (A)	19766 1.2	19120 1.2	637 0.0	18764 1.2	978 0.1	3.7	15825
25 Carinthia (A)	33322 1.0	55292 1.7	1307 0.0	42220 1.3	3428 0.1	4.1	33294
26 Styria (A)	115211 1.4	114332 1.4	2077 0.0	84320 1.0	7168 0.1	4.0	81214

F: 1892; I: 1901, * according to occupational census, ** estimated according to 1930 farm census data, Trent and Bolzano like A; CH: 1905, without farms of 0–0.5 ha; A: 1902. Explanations and sources on the next page.

Appendices

Table A.2
Based on the farm census closest to 1900; inclusion of the first Italian farm census of 1930 (in two former Austrian provinces a comparison with 1902 is possible); Liechtenstein has no official published statistics (for indications regarding its small-farming structure, see Ospelt 1972, 146). State forests in France not included; autonomous forests in Switzerland not included; for Austria, use of the standard version with all of the farms (since "pure forest economy farms," which are usually a bit smaller, account for barely 1%). The regions are adapted approximately to the administrative situation in 1990, as in table A.1.

Sources: F-Statistics 1892, Division du sol; I-Statistics 1930, vol. 2, pt. 2; CH-Statistics 1905b, Nachtrag; A-Statistics 1902, tab. 2.

Table A.3
Based on the farm census closest to 1900; for Italy, where not even the first farm census of 1930 provides workforce data, the occupational census of 1901 was used; Liechtenstein has no official published statistics (for indications regarding its small-farming structure, see Ospelt 1972, 146). Labor force: over and under 16 years (F, A); 14 years and older (CH). Administrators: *"régisseur"* (F); *"Verwalter," "Direktoren," "Angestellte,"* or *"Beamte"* (CH); *"Beamte," "Aufsichtspersonen"* (A).
The regions are adapted approximately to the administrative situation in 1990, as in table A.1.

Sources: F-Statistics 1892, Population des travailleurs agricoles; I-Statistics 1901, vol. 3; CH-Statistics 1905, tab. 4; A-Statistics 1902, tab. 20; farms as in table A.2.

Table A.4: Illegitimacy rates in Alpine regions, 1870 and 1900

Region	1870/1874		1900/1904		Change 1870–1900
	L	LS	L	LS	LS
1 Alpes-Maritimes (F)	6.0	6.3	10.8	11.6	+5.3
2 Alpes-de-Haute-Provence (F)	2.2	3.2	2.6	2.8	−0.4
3 Hautes-Alpes (F)	2.2	2.4	3.5	3.6	+1.2
4 Savoie (F)	3.7	3.8	5.2	5.5	+1.7
5 Haute-Savoie (F)	5.0	5.2	4.8	4.9	−0.3
6 Imperia (I)	5.6	5.6	5.9	6.1	+0.5
7 Val d'Aosta (I)	–	–	9.6	9.7	–
8 Sondrio (I)	3.0	3.1	3.3	3.4	+0.3
9 Trent (I/A)	–	1.1	1.2	1.3	+0.2
10 Bolzano/Bozen (I/A)	–	5.0	5.9	6.0	+1.0
11 Belluno (I)	2.8	2.8	3.1	3.1	+0.3
12 Valais (CH)	3.8	3.8	3.5	3.5	−0.3
13 Ticino (CH)	1.7	1.7	2.9	3.0	+1.3
14 Grisons (CH)	3.8	3.8	3.3	3.4	−0.4
15 Uri (CH)	2.1	2.1	1.3	1.3	−0.8
16 Unterwalden (CH)	3.4	3.5	1.3	1.4	−2.1
17 Schwyz (CH)	3.0	3.0	1.7	1.7	−1.3
18 Glarus (CH)	1.2	1.2	1.8	1.9	+0.7
19 Appenzell (CH)	3.4	3.5	2.9	2.9	−0.6
20 St. Gall (CH)	3.2	3.3	3.8	3.9	+0.6
21 Liechtenstein (FL)	–	–	–	–	–
22 Vorarlberg (A)	–	7.0	5.8	5.8	−1.2
23 Tyrol (A)	–	11.1	15.9	15.9	+4.8
24 Salzburg (A)	–	29.1	25.7	26.0	−3.1
25 Carinthia (A)	–	46.0	39.5	39.7	−6.3
26 Styria (A)	–	32.6	30.4	30.6	−2.0

L = out-of-wedlock live births as a percentage of all live births, five-year average
LS = out-of-wedlock live births and stillborns as a percentage of all births, five-year average
The regions are adapted approximately to the administrative situation in 1990, as in table A.1.

Sources: F-Statistics 1870–1874, 1900–1904; I-Statistics 1870–1874 (exposed children registered as out-of-wedlock), 1900–1904; CH-Statistics 1870–1874, 1900–1904; Gustav Adolf Schimmer, "Die unehelich Geborenen in Oesterreich 1831–1874," *Statistische Monatsschrift* 2 (1876), 168–70, A-Statistics 1900–1904. Liechtenstein has no official published statistics.

Appendices

Table A.5: Farm size, agricultural labor force, and illegitimacy rates in Alpine regions of Hapsburg Austria by district, 1870 and 1900

Region	Farm size		Labor force		Illegitimacy	
District	S	ML	LF	AS	1870	1900
9 Trent	91	3	2.7	0.1	1	1
Borgo	90	4	2.6	0.1	1	1
Cavalese	96	1	2.2	0.0	3	2
Cles	92	2	2.6	0.1	1	1
Primiero	85	4	2.5	0.0	1	1
Riva	88	4	2.8	0.1	1	1
Roveredo	86	5	2.7	0.1	1	1
Tione	93	2	2.8	0.1	1	1
Trient, city	93	1	2.9	0.6	5	5
Trient, environs	90	3	2.9	0.1	1	1
10 Bolzano/Bozen	53	32	3.4	0.8	5	6
Botzen, city	87	5	1.9	0.3	5	15
Botzen, environs	58	28	3.4	0.8	2	3
Brixen	36	48	3.5	0.9	6	6
Brunecken	40	44	3.5	0.9	6	4
Meran	61	23	3.3	0.7	8	9
22 Vorarlberg	68	14	2.3	0.2	7	6
Bludenz	57	18	2.6	0.2	8	6
Bregenz	53	25	2.5	0.2	7	7
Feldkirch	87	4	2.0	0.1	6	5
23 Tyrol	50	31	3.4	0.6	11	16
Imst	81	6	3.2	0.2	9	6
Innsbruck, city	33	44	3.9	1.0	12	26
Innsbruck, environs	53	25	3.3	0.6	8	24
Kitzbühel	28	56	3.5	1.0	24	22
Kufstein	32	52	3.8	0.9	14	14
Landeck	65	12	3.0	0.2	9	5
Lienz	24	58	4.3	1.0	8	7
Reutte	62	14	2.8	0.2	9	6
Schwaz	43	36	3.2	0.7	10	11
24 Salzburg	37	49	3.7	1.2	29	26
Salzburg, city	53	26	4.1	2.3	37	33
Salzburg, environs	39	46	3.4	0.9	23	22
St. Johann	31	56	4.0	1.4	30	22
Tamsweg	34	52	4.1	1.5	28	26
Zell am See	39	46	4.2	1.6	36	35

Table A.5 (cont'd)

Region / District	Farm size		Labor force		Illegitimacy	
	S	ML	LF	AS	1870	1900
25 Carinthia	39	47	4.1	1.3	46	40
Hermagor	41	40	3.3	0.5	27	26
Klagenfurt, city	60	20	4.0	1.8	69	58
Klagenfurt, environs	37	47	4.2	1.3	47	46
St. Veit	39	53	5.2	2.5	68	57
Spittal	37	48	3.8	1.0	39	36
Villach	49	36	3.3	0.7	36	30
Völkermarkt	33	54	4.3	1.3	40	32
Wolfsberg	33	55	4.6	1.7	39	32
26 Styria	50	33	4.0	1.0	33	31
Bruck	50	42	4.0	1.4	44	34
Deutsch-Landsberg	55	27	3.8	0.9	22	22
Feldbach	52	24	4.0	0.7	24	17
Gratz, city	90	3	3.4	1.1	47	47
Gratz, environs	48	34	4.1	1.1	26	27
Hartberg	42	40	3.9	0.8	27	16
Judenburg	42	48	4.6	2.1	45	42
Leibnitz	55	24	4.2	1.0	19	18
Leoben	68	25	3.2	1.2	43	31
Lietzen	41	45	4.0	1.4	40	38
Murau	27	62	4.9	2.2	50	48
Radkersburg	59	21	3.4	0.5	19	14
Weiz	43	37	3.9	0.9	23	19

Farm size: Small farms of up to 5 ha (= S) and mid-sized to large farms of 10 ha or more (= ML) as percentages of all farms in 1902.
Labor force: Labor force per farm (= LF) and agricultural servants per farm (= AS) in 1902.
Illegitimacy: Out-of-wedlock live births and stillborns as a percentage of all births, five-year averages, 1870–1874 and 1900–1904.
Regions are adapted approximately to the administrative situation in 1990, and with the numbers as in table A.1. Districts are indicated according to names and subdivisions of 1869.

Sources: A-Statistics 1902, tab. 2 and 20; Schimmer 1876 (see table A.4), 168–70; A-Statistics 1900–1904.

Appendices

Table A.6: Cities with 5000 or more inhabitants in the Alps, 1870 and 1900

City	Inhabitants in thousands		Annual increase	Altitude
	1869/72	1900/01	1870–1900 (‰)	(m)
Albertville (F)	4.4	6.2	11.9	345
Briançon (F)	4.2	7.4	1.5	1321
Digne (F)	6.9	7.2	3.5	608
Gap (F)	8.9	11.0	7.3	733
Grenoble (F)	42.7	68.6	16.5	214
Vizille (F)	3.9	5.0	8.6	279
Aosta (I)	7.7	7.6	−0.4	583
Belluno (I)	16.0	19.1	5.9	389
Bolzano/Bozen (I/A)	9.4	13.9	12.7	262
Bressanone/Brixen (I/A)	4.3	5.8	9.7	559
Feltre (I)	13.1	15.2	5.0	325
Lecco (I)	7.5	10.4	11.0	214
Levico (I/A)	6.3	6.3	0.0	506
Luino (I)	2.7	6.0	27.0	202
Merano/Meran (I/A)	4.2	9.3	26.0	323
Riva (I/A)	5.1	7.6	13.0	74
Rovereto (I/A)	9.1	10.2	3.7	205
Sondrio (I)	6.8	7.7	4.2	307
Susa (I)	4.4	5.0	4.3	503
Trent (I/A)	17.1	24.9	12.2	194
Maia Bassa/Untermais (I/A)	1.4	5.0	41.9	323
Dodiciville/Zwölfmalgreien (I/A)	3.3	5.3	15.4	262
Chur (CH)	7.5	11.5	14.4	596
Davos (CH)	2.0	8.1	47.7	1575
Einsiedeln (CH)	7.7	8.5	3.3	885
Glarus (CH)	5.5	4.9	−3.8	481
Lugano (CH)	5.9	9.4	15.6	271
Sion (CH)	4.9	6.0	6.8	491
Thun (CH)	4.6	6.0	8.9	562
Bludenz (A)	2.5	5.4	25.2	585
Bruck an der Mur (A)	3.8	7.6	22.6	487
Donawitz (A)	4.0	13.1	39.0	532
Fohnsdorf (A)	4.3	7.4	17.7	744
Hall (A)	5.0	6.2	7.0	579
Hallein (A)	3.6	6.6	19.7	469
Hötting (A)	3.5	5.7	15.9	574
Innsbruck (A)	16.3	26.9	16.3	574
Klagenfurt (A)	15.3	24.3	15.0	446
Knittelfeld (A)	2.0	8.1	46.2	645
Leoben (A)	4.5	10.2	26.7	532
Schwaz (A)	4.8	6.5	9.8	538
Villach (A)	4.5	9.7	25.1	501
Wilten (A)	2.8	12.5	49.4	574

Appendices

Table A.6
Cities with at least 5000 inhabitants (rounded off to the nearest hundred, thus exactly 4950 or more), with at least 3500 inhabitants in the main nucleus. Contemporary communal territories, current/contemporary national affiliations. Alpine boundaries according to morphological criteria, without the border zone at the foot of the Alps of ± 10 km from the Alpine rim. Definition of the Alps according to Dematteis 1975, 84–99.

Sources: F-Statistics 1872, tab. 10; *Paroisses et communes de France. Isère* (Paris, 1983), 709; F-Statistics 1901; I-Statistics 1871, vol. 1 and 1901, vol. 1; CH-Statistics 1870, vol. 1 and 1900, vol. 1 with 1895 and 1906; A-Statistics 1869, issue 6 with 1869b and 1900 with 1880.

Were the criteria used in table 4.1 (rounding off to the nearest 1000, thus exactly 4500 inhabitants or more, with at least 3000 in the main nucleus) to be applied here, the following cities would be added for 1870/1900: Ala (I/A) 4.2/4.9, Albino (I) 3.1/5.1, Bellinzona (CH) 2.5/4.9, Borgo (I/A) 4.8/4.4, Chiavenna (I) 4.1/4.7, Demonte (I) 7.8/7.1, Domodossola (I) 3.5/4.6, Judenburg (A) 3.2/4.9, Mezzolombardo (I/A) 3.4/4.5, Mürzzuschlag (A) 2.2/4.9, St. Veit (A) 2.3/4.7, Schwyz (CH) 6.1/7.4, Sisteron (F) 4.6/3.9.

Bibliography

Official statistical sources
Citation format: A-Statistics 1861 = *Statistisches Handbüchlein für die Oesterreichische Monarchie*. Ed. k.k. Direction der administrativen Statistik. Vienna, 1861.

A-Statistics – Austria
1861: *Statistisches Handbüchlein für die Oesterreichische Monarchie*. Ed. k.k. Direction der administrativen Statistik. Vienna, 1861.

1869: *Bevölkerung und Viehstand der im Reichsrathe vertretenen Königreiche und Länder, dann der Militärgränze. Nach der Zählung vom 31. December 1869*. Ed. k.k. Statistische Central-Commission. 6 issues. Vienna, 1871–1872.

1869b: *Orts-Repertorium der gefürsteten Grafschaft Tirol und Vorarlberg, bearbeitet auf Grundlage der Volkszählung vom 31. December 1869*. Ed. k.k. Statistische Central-Commission in Wien. Innsbruck, 1873.

1880: *Vollständiges Ortschaften-Verzeichniss der im Reichsrathe vertretenen Königreiche und Länder nach den Ergebnissen der Volkszählung vom 31. December 1880*. Ed. k.k. Statistische Central-Commission. Vienna, 1882.

1900: *Allgemeines Ortschaften-Verzeichniss der im Reichsrathe vertretenen Königreiche und Länder nach den Ergebnissen der Volkszählung vom 31. December 1900*. Ed. k.k. Statistische Central-Commission. Vienna, 1902.

Bibliography

1900b: *Oesterreichische Statistik*. Ed. k.k. Statistischen Central-Commission. *Die Ergebnisse der Volkszählung vom 31. December 1900 in den im Reichsrathe vertretenen Königreichen und Ländern*. Vols. 63–66. Vienna, 1902–1905.

1900–1904: *Bewegung der Bevölkerung der im Reichsrathe vertretenen Königreiche und Länder im Jahre 1900 (1901, 1902, 1903, 1904)*. Ed. Bureau der k. k. Statistischen Central-Commission. Vienna, 1903 (–1908).

1902: *Ergebnisse der landwirtschaftlichen Betriebszählung vom 3. Juni 1902 in den im Reichsrate vertretenen Königreichen und Ländern*. Ed. Bureau der k.k. statistischen Zentralkommision. Vienna, 1909.

1981: *Volkszählung 1981. Wohnbevölkerung nach Gemeinden (revidierte Ergebnisse) mit der Bevölkerungsentwicklung seit 1869*. Ed. Österreichisches Statistisches Zentralamt. Vienna, 1983.

CH-Statistics – Switzerland

1870: *Eidgenössische Volkszählung vom 1. December 1870*. Ed. Statistisches Bureau des eidgenössischen Departementes des Innern. 3 vols. Berne, 1872–1876.

1870–1874: *Geburten, Sterbefälle und Trauungen in der Schweiz im Jahr 1870 (1871, 1872, 1873, 1874)*. Berne, undated.

1895: *Schweizerisches Ortschaftenverzeichnis*. Ed. Eidg. statistisches Bureau. Zurich, 1895.

1900: *Die Ergebnisse der Eidgenössischen Volkszählung vom 1. Dezember 1900*. Ed. Statistisches Bureau des eidgenössischen Departementes des Innern. 4 vols. Berne, 1904–1908.

1901–1904: *Die Bewegung der Bevölkerung in der Schweiz im Jahre 1901 (1902, 1903, 1904)*. Ed. Statistisches Bureau des eidg. Departementes des Innern. Berne, 1902 (–1906).

1905: *Ergebnisse der eidg. Betriebszählung vom 9. August 1905. Vol. 2. Die Betriebe der Urproduktion*. Ed. Statistisches Bureau des eidg. Departementes des Innern. Berne, 1910.

1905b: *Ergebnisse der eidg. Betriebszählung vom 9. August 1905*. Supplement to vol. 2. Ed. Statistisches Bureau des eidg. Departementes des Innern. Berne, 1911.

1906: *Schweizerisches Ortschaftenverzeichnis*. Ed. Eidg. statistisches Bureau. Zurich, 1906.

1952: *Arealstatistik der Schweiz 1952*. Ed. Eidgenössisches Statistisches Amt. Berne, 1953.

1990: *Eidgenössische Volkszählung 1990. Bevölkerungsentwicklung 1850–1990. Die Bevölkerung der Gemeinden*. Ed. Bundesamt für Statistik. Berne, 1992.

F-Statistics – France

1862: *Statistique de la France. Agriculture. Résultats généraux de l'enquête décennale de 1862.* Strasbourg, 1968.

1870–1874: *Statistique de la France.* New ser. Vol. 1 (2, 3, 4). *Statistique annuelle.* Year 1871 (1870, 1872, 1873, 1874). Paris, 1874 (–1877).

1872: *Statistique de la France.* Second ser. Vol. 21. *Résultats généraux du dénombrement de 1872.* Paris, 1873.

1872b: *Statistique de la France. Résultats généraux du dénombrement de 1872.* Nancy, 1874.

1892: *Statistique agricole de la France.* Ed. Ministère de l'Agriculture. *Résultats généraux de l'enquête décennale de 1892.* Paris, 1897.

1900–1904: *Statistique annuelle du mouvement de la population pour les années 1899 et 1900 (1901, 1902, 1903, 1904). Statistique générale de la France.* Vols. 29 and 30 (–34). Paris, 1901 (–1906).

1901: Ministère de l'Intérieur. *Dénombrement de la population 1901.* Paris, 1902.

1901b: Ministère du commerce, de l'agriculture, des postes et des télégraphes. *Résultats statistiques du recensement général de la population effectué le 24 mars 1901.* 5 vols. Paris, 1904–1907.

1988: INSEE Institut National de la Statistique et des Etudes Economiques. *Inventaire Communal 1988.* Printout of computerized database.

1990: INSEE Institut National de la Statistique et des Etudes Economiques. *Recensement général de la population 1990. Population légale.* Departmental issue. Paris, undated.

FL-Statistics – Principality of Liechtenstein

1962: Fürstentum Liechtenstein. *Wohnbevölkerung - Volkszählungen 1812–1930.* Ed. Amt für Statistik. Vaduz, 1962.

1991: *Statistisches Jahrbuch 1991 Fürstentum Liechtenstein.* Ed. Amt für Volkswirtschaft. Vaduz, 1991.

I-Statistics – Italy

1870–1874: *Movimento dello stato civile nell'anno 1870 (1871, 1872, 1873, 1874).* Ed. Ministero di Agricultura, Industria e Commercio. Milan (Florence, Rome), 1872 (–1876).

1871: Ministero di Agricultura, Industria e Commercio. *Censimento 31 dicembre 1871.* 3 vols. Rome, 1874–1876.

Bibliography

1900–1904: Ministero di Agricultura, Industria e Commercio. *Statistica della popolazione. Movimento dello stato civile. Anno 1900 (1901, 1902, 1903, 1904)*. Rome, 1902 (–1906).

1901: Ministero di Agricultura, Industria e Commercio. *Censimento della popolazione del Regno d'Italia al 10 febbraio 1901*. 5 vols. Rome, 1902–1904.

1930: Istituto Centrale di Statistica del Regno d'Italia. *Censimento generale dell'agricoltura, 19 Marzo 1930–VIII*. Vol. 2. *Censimento delle aziende agricole*. 2 parts. Rome, 1935–1936.

1985: ISTAT Istituto nazionale di statistica. *Popolazione residente e presente dei comuni. Censimenti dal 1861 al 1981*. Rome, 1985.

1988: ISTAT Istituto nazionale di statistica. *Comuni, comunità montane, regioni agrarie al 31 dicembre 1988. Codici e dati strutturali*. Rome, 1990.

1991: *Südtirol-Handbuch*. Ed. Südtiroler Landesregierung. Bozen, 1991.

Selected literature

This list is comprised of a selection of regional and general studies relating to the history of the Alpine space and to key themes discussed. Complete references to other works used for the study are provided in the notes.

Albera, Dionigi. *L'organisation domestique dans l'espace alpin. Equilibres écologiques, effets de frontières, transformations historiques*. Thèse d'ethnologie (PhD dissertation). Université de Provence, 1995 (forthcoming).

Atlas zur Geschichte des steirischen Bauerntums. Graz, 1976.

Bairoch, Paul and Jean Batou, Pierre Chèvre. *La population des villes européennes de 800 à 1850 / The Population of European Cities from 800 to 1850*. Geneva, 1988.

Baratier, Édouard (ed.) *Histoire de la Provence*. 2 vols. Toulouse, 1978.

Bätzing, Werner et al. *Der sozio-ökonomische Strukturwandel des Alpenraumes im 20. Jahrhundert. Eine Analyse von "Entwicklungstypen" auf Gemeinde-Ebene im Kontext der europäischen Tertiarisierung*. Berne, 1993.

Bätzing, Werner. *Die Alpen. Entstehung und Gefährdung einer europäischen Kulturlandschaft*. Munich, 1991.

Bergier, Jean-François. *Pour une histoire des Alpes, Moyen Age et Temps modernes*. Hampshire, 1997.

Bergier, Jean-François and Sandro Guzzi (ed.) *La découverte des Alpes*. Basel, 1992.

Bilgeri, Benedikt. *Geschichte Vorarlbergs*. 5 vols. Vienna, 1971–1987.

Bircher, Ralph. *Wirtschaft und Lebenshaltung im schweizerischen Hirtenland am Ende des 18. Jahrhunderts.* Lachen, 1938.
Blanchard, Raoul. *Les Alpes Occidentales.* 7 vols. Grenoble, 1938–1956.
Bligny, Bernard (ed.) *Histoire du Dauphiné.* Toulouse, 1973.
Blockmans, Wim and Jean-Philippe Genet (ed.) *Visions sur le développement des états européens. Théories et historiographies de l'état moderne.* Rome, 1993.
Blum, Jerome. *The End of the Old Order in Rural Europe.* Princeton, 1978.
Boserup, Ester. *Population and Technology.* Oxford, 1981.
Boserup, Ester. *The Conditions of Agricultural Growth. The Economics of Agrarian Change Under Population Pressure.* London, 1993 (1965).
Braudel, Fernand. *The Mediterranean and the Mediterranean World in the Age of Philip II.* 2 vols. Trans. Siân Reynolds. London, 1972. First French ed. 1949, expanded 1966.
Brönnimann, Stefan. "Die schiff- und flössbaren Gewässer in den Alpen von 1500 bis 1800. Versuch eines Inventars." *Der Geschichtsfreund* 150 (1997): 119–78.
Carlen, Louis and Gabriel Imboden (ed.) *Alpe und Alm. Beiträge zur Kulturgeschichte des Alpwesens.* Brig, 1994.
Cavaciocchi, Simonetta (ed.) *Le migrazioni in Europa secc. XIII–XVIII.* Florence, 1994.
Ceschi, Raffaello. "Migrazioni dalla montagna alla montagna." *Gewerbliche Migration im Alpenraum. Historikertagung in Davos 25.–27. IX. 1991.* Bozen/Bolzano, 1994. Pp. 15–45.
Chittolini, Giorgio. "Principe e comunità alpine in area lombarda alla fine del medioevo." In Martinengo 1988. Pp. 219–35.
Cole, John W. and Eric R. Wolf. *The Hidden Frontier. Ecology and Ethnicity in an Alpine Valley [Trentino-Südtirol].* New York, 1974.
Coppola, Gauro and Pierangelo Schiera (ed.) *Lo spazio alpino: area di civiltà, regione cerniera.* Naples, 1991.
Coppola, Gauro. "La montagna alpina. Vocazioni originarie e trasformazioni funzionali." In *Storia dell'agricoltura italiana in età contemopranea.* Ed. Piero Bevilacqua. Vol. 1. Venice, 1989. Pp. 495–530.
Cotrao (ed.) *L'homme et les Alpes.* Grenoble, 1992.
Dainelli, Giotto. *Le Alpi.* 2 vols. Turin, 1963.
Dematteis, Giuseppe. "Le città alpine." In *Le città alpine. Documenti e note.* Ed. Bruno Parisi. Milan, 1975. Pp. 5–103.

Bibliography

Des Alpes traversées aux Alpes vécues / Vom Alpenübergang zum Alpenraum [on the state of the research]. *Histoire des Alpes* 1 (1996).

Dubuis, Pierre. "Les hommes et le milieu montagnard dans l'histoire européenne." In *Ninth International Economic History Congress Bern 1986. Debates and Controversies.* Zurich, 1986. Pp. 3–19.

Favier, René. *Les villes du Dauphiné aux XVIIe et XVIIIe siècles.* Grenoble, 1993.

Fontaine, Laurence. *Histoire du colportage en Europe (XVe–XIXe siècles).* Paris, 1993.

Fräss-Ehrfeld, Claudia. *Geschichte Kärntens.* 3 vols. (so far). Klagenfurt, 1984/1994/2000.

Frödin, John. *Zentraleuropas Alpwirtschaft.* 2 vols. Oslo, 1940–1941.

Gerosa, Pier Giorgio. "La città delle Alpi nella storiografia urbana recente." In Martinengo 1988. Pp. 139–59.

"Geschichte der Land- und Forstwirtschaft." In *Wirtschafts- und Sozialgeschichte der Slowenen, nach Wirtschaftszweigen enzyklopädisch behandelt (Gospodarska in druzbena zgodovina Slovencev. Enciklopedicna obravnava po panogah).* Ed. Izdaja Slovenska akademija znanosti in umetnosti. 2 vols. Ljubljana, 1970–80. German summary in vol. 2, pp. 555–655.

Geschichte Salzburgs. Stadt und Land. Ed. Heinz Dopsch and Hans Spatzenegger. 2 vols in 3 and 5 pts. Salzburg, 1983–1991.

Guichonnet, Paul (ed.) *Histoire et Civilisations des Alpes.* 2 vols. Toulouse and Lausanne, 1980.

Guichonnet, Paul (ed.) *Nouvelle histoire de la Savoie.* Toulouse, 1996.

Gunst, Peter und Tamás Hoffmann (ed.) *Grand domaine et petites exploitations en Europe au moyen âge et dans les temps modernes. Rapports nationaux.* Budapest, 1982.

Guzzi, Sandro. *Agricoltura e società nel Mendrisiotto del Settecento.* Bellinzona, 1990.

Handbuch der Bündner Geschichte. Ed. Verein für Bündner Kulturforschung. 4 vols. Chur, 2000.

Histoire des Alpes - Storia delle Alpi - Geschichte der Alpen (Annual Journal of the International Association for Alpine History). Zurich, 1996–.

Histoire du Canton de Fribourg. 2 vols. Fribourg, 1981.

Histoire et actualité de la transhumance en Provence, Les Alpes de Lumière 95/96. Forcalquier, 1986.

Holenstein, André. "Bauern zwischen Bauernkrieg und Dreissigjährigem Krieg." *Enzyklopädie deutscher Geschichte* 38. Munich, 1996.

Janin, Bernard. *Une région alpine originale. Le Val d'Aoste. Tradition et renouveau.* Grenoble, 1968.

Kain, Roger J. P. and Elizabeth Baigent. *The Cadastral Map in the Service of the State. A History of Property Mapping.* Chicago and London, 1992.

Klein, Kurt. "Die Bevölkerung Österreichs vom Beginn des 16. bis zur Mitte des 18. Jahrhunderts." In *Beiträge zur Bevölkerungs- und Sozialgeschichte Österreichs.* Ed. Heimold Helczmanovszki. Vienna, 1973. Pp. 47–112.

Körner, Martin and François Walter (ed.) *Quand la Montagne aussi a une Histoire. Mélanges offerts à Jean-François Bergier.* Berne, 1996.

Le Alpi e l'Europa. Atti del convegno di studi Milano 4–9 ottobre 1973. 5 vols. Bari, 1974–1977.

L'élevage et la vie pastorale dans les montagnes de l'Europe au moyen âge et à l'époque moderne. Clermont-Ferrand, 1984.

Les Alpes de Slovénie / Die Alpen Sloweniens. Histoire des Alpes 2 (1997).

Livi Bacci, Massimo. "La ricostruzione del passato: dall'individuo alla collettività." In Philippe Braunstein et al. *Il mestiere dello storico dell'Età moderna. La vita economica nei secoli XVI–XVIII.* Bellinzona, 1997. Pp. 139–54.

Martinengo, Edoardo (ed.) *Le Alpi per l'Europa. Una proposta politica. Economia, territorio e società. Istituzioni, politica e società.* Milano, 1988.

Martonne, Emmanuel de. *Les Alpes. Géographie générale.* Paris, 1926.

Mathieu, Jon. "'Ihre Geschichte besteht darin, keine zu haben.' Die Alpen der frühen Neuzeit im Spannungsfeld wissenschaftlicher Disziplinen." In *Die Frühe Neuzeit in der Geschichtswissenschaft. Forschungstendenzen und Forschungserträge.* Ed. Nada Boskovska Leimgruber. Paderborn, 1997. Pp. 109–26.

Mathieu, Jon. *Eine Agrargeschichte der inneren Alpen. Graubünden, Tessin, Wallis 1500–1800.* Zurich, 1992.

Mathis, Franz. *Zur Bevölkerungsstruktur österreichischer Städte im 17. Jahrhundert.* Munich, 1977.

Mattmüller, Markus (ed.) *Wirtschaft und Gesellschaft in Berggebieten.* Basel, 1986.

Mitterauer, Michael. "Formen ländlicher Familienwirtschaft. Historische Ökotypen und familiale Arbeitsorganisation im österreichischen Raum." In *Familienstruktur und Arbeitsorganisation in ländlichen Gesellschaften.* Ed. Josef Ehmer, Michael Mitterauer. Vienna, 1986. Pp. 185–324.

Bibliography

Mitterauer, Michael. *Historisch-anthropologische Familienforschung. Fragestellungen und Zugangsweisen.* Vienna and Cologne, 1990.

Mobilité spatiale et frontières / Räumliche Mobilität und Grenzen. Histoire des Alpes 3 (1998).

Mozzarelli, Cesare and Giuseppe Olmi (ed.) *Il Trentino nel Settecento fra Sacro Romano Impero e antichi stati italiani.* Bologna, 1985.

Netting, Robert McC. *Balancing on an Alp. Ecological Change and Continuity in a Swiss Mountain Community* [Valais]. Cambridge, 1981.

Netting, Robert McC. *Smallholders, Householders. Farm Families and the Ecology of Intensive, Sustainable Agriculture.* Stanford, 1993.

Nicolas, Jean. *La Révolution Française dans les Alpes. Dauphiné et Savoie 1789–1799.* Toulouse, 1989.

Nicolas, Jean. *La Savoie au 18e siècle. Noblesse et bourgeoisie.* 2 vols. Paris, 1978.

Niederer, Arnold. *Alpine Alltagskultur zwischen Beharrung und Wandel. Ausgewählte Arbeiten aus den Jahren 1956 bis 1991.* Berne, 1993.

Ospelt, Alois. "Wirtschaftsgeschichte des Fürstentums Liechtenstein im 19. Jahrhundert. Von den napoleonischen Kriegen bis zum Ausbruch des Ersten Weltkrieges." In *Jahrbuch des Historischen Vereins für das Fürstentum Liechtenstein* 72 (1972): 5–423 and appendix pp. 1–267.

Peters, Jan (ed.) *Gutsherrschaft als soziales Modell. Vergleichende Betrachtungen zur Funktionsweise frühneuzeitlicher Agrargesellschaften (Historische Zeitschrift Beiheft NF 18).* Munich, 1995.

Pfister, Christian. *Im Strom der Modernisierung. Bevölkerung, Wirtschaft und Umwelt im Kanton Bern 1700–1914.* Berne, 1995.

Pickl, Othmar. "Brandwirtschaft und Umwelt seit der Besiedlung der Ostalpenländer." In *Wirtschaftsentwicklung und Umweltbeeinflussung (14.–20. Jahrhundert).* Ed. Hermann Kellenbenz. Wiesbaden, 1982. Pp. 27–55.

Rebel, Hermann. *Peasant Classes. The Bureaucratization of Property and Family Relations under Early Habsburg Absolutism 1511–1636* [Upper Austria]. Princeton, 1983.

Rodger, Richard (ed.) *European Urban History. Prospect and Retrospect.* Leicester and London, 1993.

Rosenberg, Harriet G. *A Negotiated World: Three Centuries of Change in a French Alpine Community* [Hautes-Alpes]. Toronto, 1988.

Sandgruber, Roman. *Österreichische Agrarstatistik 1750–1918.* Munich, 1978.

Scaramellini, Guglielmo. *La Valtellina fra il XVIII e il XIX secolo. Ricerca di geografia storica.* Turin, 1978.

Schiera, Pierangelo et al. (ed.) *L'autonomia e l'amministrazione locale nell'area alpina.* Milan, 1988.

Siddle, David J. "Inheritance Strategies and Lineage Development in Peasant Society [Savoy]." *Continuity and Change* 1 (1986): 333–61.

Stolz, Otto. *Rechtsgeschichte des Bauernstandes und der Landwirtschaft in Tirol und Vorarlberg.* Bozen, 1949.

Tilly, Charles. *Coercion, Capital, and European States, AD 990–1992.* Oxford 1992.

Vendramini, Ferruccio. *Le comunità rurali bellunesi (secoli XV e XVI).* Belluno, 1979.

Viallet, Hélène. *Les alpages et la vie d'une communauté montagnarde: Beaufort du Moyen Age au XVIIIe siècle.* Annecy, 1993.

Viazzo, Pier Paolo. *Upland Communities. Environment, Population and Social Structure in the Alps Since the Sixteenth Century.* Cambridge, 1989.

Westermann, Ekkehard (ed.) *Internationaler Ochsenhandel (1350–1750).* Stuttgart, 1979.

Index

Abel, Wilhelm 112, 121
adaptation, theory of 128–31, 225
agararian dualism (Agrardualismus) 200–201
agrarian crisis 56, 98, 128
agrarian intensification 49–50, 53, 58, 60-70, 75, 77, 98, 102–3, 109, 116–18, 122–24, 130–31, 146, 147, 156, 197, 204, 210, 211, 214, 222–23, 225–26
agrarian output 48, 52, 57, 58, 60, 64, 66, 68, 76–77, 213–14
agrarian reform 148–53, 156
agrarian servants. See hired hands.
agrarian structure (Agrarverfassung) 8–9, 48, 135–37, 139, 144–45, 146, 154, 156–57, 161, 180, 188, 195–97, 199, 200, 203–5, 207, 209, 224
agrarian surplus 97
agricultural workers 39, 101–2, 107, 116, 129, 143, 146, 157, 187, 230

agriculture. See agrarian crisis, agrarian dualism, agrarian intensification, agrarian production, agrarian reform, agrarian structure, agrarian surplus, and specific keywords.
Aix-en-Provence 15
Alagna 170, 178
Albera, Dionigi 125, 162–63, 170, 179, 189
Allèves illustration 7
Alpes-de-Haute-Provence 20, 28, 36–37, 40, 41, 59, 67, 139, 230–34
Alpes-Maritimes 28, 37, 67, 74, 143, 144, 153, 230–34
Alpiculture illustration 12, 47, 49–55, 76, 121, 131
Alpine Convention 5, 6, 25
Alpine economy. See Alpiculture.
Alpine intensity differential 49–55, 76–77, 97, 117, 122, 228
Alpine pasture workers 52, 143

250

Index

Alpine regionalism   6, 7, 195

ALPS ADRIATIC working community   5, 6

Alps, boundaries   9–10, 25–26, 85–86, 116

Alps, central   66, 72, 199, 213, 224, 227

Alps, eastern   56, 72, 139, 141–42, 146, 155, 211, 217, 217, 224

Alps, ideology   6, 84, 227–28

Alps, internal differences   35–36, 39–40, 55, 60–61, 72, 117, 118, 144–45, 157, 162–63, 189–90, 216–17, 223

Alps, northern slope   57, 61, 63, 68, 122

Alps, research   3, 12–14

Alps, situation and relief   illustration 1, 9–11, 84

Alps, southern   36, 51, 59, 62–63, 64–65, 107, 170

Alps, southern slope   59, 64–65, 68, 119, 125, 139, 156, 199

Alps, western   28, 41–42, 59, 60, 72, 128, 139, 170, 211, 213, 224, 227

altitudinal location   35, 39, 40–43, 52, 53, 55–56, 57, 59–61, 66, 70, 73, 77, 92, 98, 99, 102–3, 107–9, 116–18, 128, 169, 223, 226

Alto Adige/Südtirol   20, 50, 57, 65, 69, 144, 152–53, 163. *See also* Bolzano/Bozen

Ampezzo   34

Amsterdam   228

Anderson, Perry   136

animal husbandry   50, 55, 61–64, 69, 70, 76, 121, 130

Annecy   89

anthropology   13–14, 48, 127, 162, 213

Aosta   36, 83, 89, 99, 237. *See also* Val d'Aosta

Appennines   9, 86

Appenzell   28, 37, 39, 67, 230–34

ARGE ALP working community   5, 6, 20

artisanship (industry, commerce)   36, 39, 95–97, 125, 185

Asiago   85

Augsburg   95, 119

Austria   5, 6, 16, 17, 20, 25, 33, 35, 38, 40, 41, 48, 65, 66, 69, 92, 94, 106, 120, 129, 137–39, 143–46, 151–56, 161–63, 181, 184, 188, 198, 200, 210, 217, 235

Austria, Lower   17, 50, 60, 149, 152, 202, 203, 206

Austria, Upper   17, 50, 60, 149, 203, 205, 206, 216, 224

autonomy   7–8, 16, 19, 95, 127, 183, 199, 212–14, 217

Avance valley   147

Avançon   147

Bad Ischl. *See* Ischl.

Bairoch, Paul   85–86, 90–91, 97

Barcelonnette   85

Basel   204

Bassano   89

Bätzing, Werner   7, 25, 32, 34, 42, 116

Bavaria   6, 17, 143, 149, 204, 205, 216, 224

Beaufortin   62, 64, 122

Bellinzona   93, 238

Bellunese, province of Belluno   28, 33–34, 37, 40, 62, 65, 67, 73, 143, 153, 230–32, 234

Index

Belluno 89, 101, 237
Berchtesgaden 5
Bergamo 89
Bergier, Jean-François 13, 22, 25
Berne 16, 42, 60,
Biella 89
Blanchard, Raoul 42, 128–30
Bobek, Hans 98
Bohemia 198, 203
Bolzano/Bozen 56, 89, 92, 99, 111, 112, 235, 237
Bolzano/Bozen, province of 6, 28, 37, 67, 138, 145, 230–32, 234–35. *See also* Alto Adige
Bonaparte, Napoleon 149, 154, 215
Bonnin, Bernard 214
borders, density 6, 110, 224–25
borders, formation 15, 19, 107–8
borders, theory 19–20
Bormio 172
Boserup, Ester 78, 97, 196
Botero, Giovanni 83
bourgeoisie 151, 196, 201, 227. *See also* state - citizens, intellectuals
bovines 50, 52, 55, 61–63. *See also* livestock trade
Braudel, Fernand 12, 13, 123, 196, 216
Bregenz 143
Brenner pass 20, 92–93
Brescia 89
Briançon 103, 108, 147, 213, 237
Briançonnais 103
Bruck an der Mur 107, 237
buckwheat 59, 65
Burgundy 164
butter. *See* cheese production.

cadastre illustration 7, 166–68, 172, 188, 198–99, 209, 217
Carinthia 6, 17, 28, 37, 50, 51, 65–67, 72, 75, 95, 122, 143–45, 149, 152, 155, 161, 180–90, 206, 208, 214, 223, 230–32, 234, 236
Carniola 17, 152, 181
Caroni, Pio 178
Casteldelfino 170, 178
Cateau-Cambrésis 163
centers, formation of 17, 94–97, 103, 117, 179, 182
cereals 49, 52, 55, 58, 60, 65, 68, 77, 101, 106, 122
Chambéry 16, 89, 164–65, 167–68, 179
cheese production 51, 52, 62, 70, 101, 122–23
chestnuts 52, 55, 64
Chevaline 169, 178
Chiavenna 172, 238
Chittolini, Giorgio 198, 199
Chur 17, 19, 92, 172, 173, 179, 237
Chur, bishop of 17, 173, 174
church. *See* tithes, clerics, Protestant reform.
CIPRA (International Commission for the Protection of the Alps) 6
Cisalpine Republic 149, 172
cities, concept of 83–84, 104, 171–72
cities, density 85, 86, 97, 108–10, 119, 122, 201
cities, growth 86, 91, 92–97, 104–6, 109, 116–18, 179, 197, 223
cities, provisioning of 97–102, 106, 108, 109, 121
cities, size 85–92, 99, 104–5, 237–38

252

Index

citizens. *See* state.

civil code   149, 154, 179

clerics   93, 95, 170, 174, 176

clientelism   175, 212

climate, history of   56

clover, forage grass   50, 57, 66

commune, community   14, 163, 166, 171–76, 206, 213, 223, 227

Como   89, 125, 177

Coppola, Gauro   198

corn/maize   illustration 4, 50, 55, 59, 64–68, 77, 116, 223

corvée labor (forced labor)   151, 152, 196, 200, 202–3

COTRAO, Western Alps Working Community   5, 6

cow-men   63

cows. *See* bovines.

Cremona   119

Croatia   6, 203

cropping frequency   55–61, 116, 223

cultivation, multi-cropping   55–56, 59

cultivation, swidden   58

cultivation, with hoes   75

culture, cultural factors   3, 101, 126, 170, 175–76, 177, 210

Cuneo   89

Dauphiné   15, 19, 41, 64, 74, 96, 103, 147, 214

Davos   105, 108, 117, 172, 237

day laborers   101, 143, 145, 186, 210, 232

Dematteis, Giuseppe   84, 86, 116

democracy   136, 147, 216

demographic crisis   63, 199, 203. *See also* epidemics

demography   14, 23–24, 43, 128, 131, 225–26

Derouet, Bernard   210

domestic unit   129, 162–63, 169–70, 177–79, 185–87, 189–90, 206, 208, 212, 224. *See also* farming establishment

Donawitz   107, 237

Dronero   89

Dürer, Albrecht   illustration 2

ecology   127–31, 205, 210

Emanuel Filibert of Savoy   16, 163–66, 168

emigration. *See* migration.

emperor, German/Austrian   149, 151, 154

emphyteusis   148, 174, 179–80, 199

Engadine   59, 73, 176–77

Enlightenment   214–17

ennoblement   167, 175, 183

environment   12, 48, 53, 63, 70–73, 77, 102, 114–15, 118, 127–31, 209–10, 215, 222, 225. *See also* climate history, altitude, steepness, vegetation

epidemics   36, 124

estates, provincial / social orders   95, 96, 147, 151, 164, 181–82, 198–99, 202

Europe   7, 9, 13, 15, 35, 39, 41, 66, 85, 94, 97, 110, 121–22, 125, 136, 138, 144, 149–50, 151, 162, 164, 173, 190, 196, 200–201, 204, 205, 216, 228,

Europe, integration of   6

Exilles   57

fallow   49, 59–60, 102

family. *See* domestic unit.

253

Index

farming establishment   137–46, 157, 197, 205–9, 212, 216, 223, 231, 235–36. *See also* domestic unit

Favier, René   41, 96

Ferdinand I, emperor   17

Ferdinand II, emperor   183

fertilizer   56–57, 59, 60, 102

feudal obligations   150–53, 167, 199–200

feudal rights   illustration 8, 150, 167–68, 180, 197

feudal rights, redemption of   148, 152–53, 154, 157, 172, 198

feudal, feudalism   148, 157, 190, 200, 213, 216, 227. *See also* feudal rights, feudal obligations

Foderé, François Emmanuel   74

Fohnsdorf   107, 237

Fontaine, Laurence   123

forage   49–53, 56–58, 60, 62, 64, 71–73, 98, 116, 118

forests   49, 55, 58, 119–21, 128, 139, 145, 156

France   5–7, 16, 20, 35, 38, 63, 106, 114, 123, 137, 139, 148–50, 154, 156, 163–64, 172, 200, 210, 217

France, king of   96, 147, 163, 174, 206

Franz Joseph I, emperor   152

freedom/s, liberty   214–17

Freistift (free, reversible tenancy)   180, 188

French Revolution   16, 33, 146–49, 151, 154, 161, 170, 178, 215

Fribourg (Switzerland)   61, 62, 69–70

Friesland   227

Friuli   162, 81

Frödin, John   48, 51, 53

fruit production   52, 64, 68–69, 102

Fugger family   95

Gap   89, 101, 102, 147, 237

Genoa   16

geography (as a discipline)   7, 13, 48, 51, 56, 83, 84, 98, 115, 127, 195

Germany   5–6, 25, 35, 38, 114, 200, 204–5, 217,

Germany, southwestern   204–5

Glarus   28, 37, 39, 67, 107, 230–32, 234, 237

goats   62, 63–64

Gorizia   89

Grasse   89

Graz   17, 65, 89

Grenoble   15, 19, 89, 92, 96, 102, 103, 107, 108, 116, 147, 214, 237

Grisons   6, 17, 28, 37, 51–53, 67, 72, 73, 105, 125, 154, 161, 171–80, 165, 185–87, 189–90, 206, 223, 230–32, 234

growth. *See* cities, growth; growth differentials; agrarian intensification; population - growth.

growth differentials   41–43, 115–18, 125

Grundherrschaft. *See* lordship, tribute-based.

Guichonnet, Paul   13, 24–25

Gurktaler Alps   185, 186

Gutsherrschaft. *See* lordship, estate-based.

Hall   93–94, 237

Haller, Albrecht von   215

Hannibal   114

254

Index

Hapsburg, house of 16, 17, 50, 56, 60, 95, 96, 108, 137, 138, 143, 151, 152, 156, 161, 174, 180, 181–84, 198, 202, 208, 215

Hautes-Alpes 28, 36, 37, 38, 67, 76, 120–21, 139, 147, 230–32, 234

Haute-Savoie 28, 37, 38, 59, 67, 161, 230–32, 234

hay. *See* forage.

Helvetic Confederation 16, 17, 19, 20, 96, 136, 149, 150. *See also* Switzerland

Helvetic Republic 150

hired hands (Dienstboten) 101, 129, 141, 143–44, 146, 157, 186, 210–11, 232–33, 235–36

Hirtenland (pastoral region) 69–70

history (as a discipline) 8, 11–12, 23–24, 47–48, 84, 172–73

Hochosterwitz illustration 9

Hötting 105, 237

household head 157, 169, 171, 186, 212

Hungary 121, 201, 203

Hüttau illustration 13

Ilanz 172

illegitimacy 142, 144–46, 157, 187, 234–36

Illyria 149

Imperia, province 28, 37, 39, 40, 67, 143, 230–32, 234

industrialization 39, 75–76, 96, 107, 123

inheritance 154–55, 162–63, 167, 169, 171, 172, 176–79, 189, 204–12, 223–24, 226

inheritance law. *See* inheritance.

Inn valley 56, 69

Innsbruck 6, 17, 19, 20, 86, 89, 92, 93, 96, 98–100, 105, 121, 196, 235, 237

intellectuals 114, 151, 155, 215, 217, 228

Inwohner 185–86

irrigation 50, 57, 58, 77, 117

Irsigler, Franz 204

Ischl 108

Italy 5, 6, 9, 16, 20, 33, 35, 38, 39, 48–50, 59, 63, 65, 86, 91, 114, 119, 122, 137, 164, 170, 197, 198–99, 201, 217

Italy, northern 49–50, 63, 86, 91, 122, 170, 197–99, 201. *See also* Lombardy, Piedmont, Veneto

Ivrea 89

Johann of Austria, archduke 215

Joseph II, emperor 151, 188

justice 96, 151–52, 164–65, 170, 174, 175, 177–79, 182–83, 208

Kant, Immanuel 215

Kempten 89, 99

kin groups 162, 169, 193, 212

Klagenfurt 17, 89, 92, 95, 99, 180, 182, 236, 237

Klein, Kurt 33

Knittelfeld 105, 106, 237

Knittler, Herbert 202–3

Laas 57

labor force. *See* hired hands, Inwohner, day laborers, Alpine pasture workers.

Lana 56

land improvement, water control projects 49–50, 60–61, 76–77, 117, 130

255

Index

land use, farming   49–50, 79
land, unproductive   35, 138–39
land-based dues   148, 150, 174
law, formation of   205–9, 212, 223–24
Lech valley   128
Lefebvre, Georges   147
legal codification   154–55, 175–76, 205
Leoben   107, 236–37
Leopold I, emperor   illustration 5
Lepetit, Bernard   33
Lesdiguières, duke of   96
Liechtenstein   5, 28, 37, 67, 217, 230–32, 234
Ligurian Republic   149
Linz   17
Livi Bacci, Massimo   23
Ljubljana   17
Lombardy   6, 16, 17, 50, 122, 125, 149, 167, 199
lordship (Herrschaft). See feudal rights, justice, household head, tribute-based lordship, estate-based lordship, serfdom, subjects.
lordship, estate-based   196, 200, 202–4, 207
lordship, tribute-based   152–53, 154, 156, 168, 174–75, 182–83, 188–89, 196, 200, 202–3, 206, 213–14, 227
Lucerne   89, 93, 215
Lugano   5, 122, 237
Lyon   126

Mainz   204
Malthus, Thomas R.   47, 53
Manosque   89
Maria Theresa, empress   188, 209

Maribor   89
market, formation of   69, 94, 122, 124, 126, 149, 153, 197, 201
market, on a wide scale   69, 70, 95, 96, 106, 119–23
market, urban   93, 97–99, 101–2
marriage   124, 128, 163–64, 168–69, 175, 177
Martonne, Emmanuel de   54, 55, 58, 86
Mathis, Franz   93
Matthiolus, Petrus Andreas   illustration 4
Mattmüller, Markus   42
Maurienne   62
Maximilian I, emperor   17
meadow cultivation. See forage.
Medicus, Ludwig Wallrath   50–51
men. See sex.
Merano   56, 237
Mercier, Louis-Sébastien   215
Meyer, Therese   186
migration   14, 36, 63, 110, 119, 123–25, 127, 128, 226
Milan   6, 16, 86, 119, 122, 174, 177
Milan, duchy   198–99. See also Lombardy
military, militarization   15, 19, 96, 103, 107, 149–50, 164, 166, 172, 181, 184, 190, 201–2
mining industry   94–95, 107
Mitterauer, Michael   94, 129–30, 210
Moldavia, principality   121
Monaco   5, 217
Mondovì   89
monetization. See market, formation of.
Mont Blanc   9

Mont Cenis 92
Montdauphin 103
Montgenèvre 92, 206
Montmin 169, 178
Morard, Nicolas 61
Moravia 200, 203
mountain climbing 61
mules 72
Munich 17
Münster, Sebastian 58
Mur valley 107
Murau, district 145
nature. *See* environment.
neighborhood 155–56. *See also* commune
Nendaz illustration 12
Netting, Robert McC. 128–29
Nice 7, 149, 164
Nicolas, Jean 167
Niederer, Arnold 13
nobility 93, 95, 96, 147, 151, 167, 168, 169, 171, 173, 175, 180, 183, 184, 188, 189, 200, 202, 203, 205, 206, 208, 213, 214
notaries 166, 168, 170–71, 176, 177–78, 189
offices. *See* administration.
Oisans 148
overpopulation 43, 123–24, 128

Paradeiser, Augustin 180
Paris 147, 215
pass-state 20
pastures 49–55, 63, 121, 145, 155
Pavia 119

peasants, emancipation of. *See* feudal rights, redemption of.
peasants, largeholder 136, 139–40, 143, 157, 204, 205, 208, 210–11, 223
peasants, smallholder 135–36, 139–40, 143, 157, 204, 205, 206, 210–11, 223
pension 185–86
periphery. *See* centers, formation of; borders, formation.
Peyer, Hans Conrad 61
Pickl, Othmar 121
Piedmont 16, 41–42, 57, 64, 72, 83, 122, 124, 125, 126, 149, 162, 164, 165, 167, 168, 170, 199
Pinerolo 89
plain, areas surrounding the Alps 6, 9, 11, 13, 15, 17, 19, 20, 34, 36, 38, 41, 43, 49–50, 51, 55–58, 66, 74, 84–88, 91, 97, 99, 109–10, 115–27, 131, 136, 205, 207, 214, 216, 222–24, 226, 228
plow 74–75
Pontechianale 170, 178
population 26, 28, 34–35. *See also* demographic crisis, overpopulation
population, density 35–41, 43, 53, 55, 63, 91, 92, 97, 99, 106, 109, 118–20, 121, 124, 126, 127, 128, 198, 216, 223, 226, 227
population, growth 34–38, 58–59, 70, 77, 104–6, 116, 120, 125, 130, 197, 222
Pordenone, province 42
Poschiavo valley 179
possession. *See* property.
potato 49, 55, 64, 66–68, 70, 77, 116, 129, 223
Preintal illustration 10, 120

Index

prices 106, 122, 123, 132, 147, 150
productivity 35, 48, 78, 210. *See also* workload, agrarian output
property, concept 188–89, 197–98
property, formation of 149, 152, 167, 197
property, forms of 136–37, 139, 143, 155–56
Protestant reform 174, 183–84
Provence 15, 59, 62, 63, 147, 148, 162
public administration 95, 96, 103, 104, 107–8, 120, 127, 138, 164–65, 166, 170–71, 181

railroads 13, 93, 107, 108
Ratzel, Friedrich 122
rent, renter 101, 153, 197, 198
republic 148, 171, 216
resources 8, 12, 14, 47, 48, 50, 52, 53, 58, 64, 68, 70, 94, 120, 123, 124, 128, 130–31, 154–57, 223, 225
revolt, revolution 95, 147, 150, 151–53, 156, 179, 215
Rhine valley 61, 68, 69, 200, 201, 204
Rhône-Alps, region 6
rice 50
river valleys, "flatland bays" ("golfes de plaine") 55, 61, 68, 76–77, 86, 117, 130, 156
roads, construction 72, 73, 106
Robot. *See* corvée.
Rosenberg, Harriet G. 121
Rousseau, Jean-Jacques 50, 228
Rovereto 89, 92, 99, 237
rural population. *See* agricultural workers, peasants, hired hands, Inwohner, day laborers, Alpine pasture workers.

Sabean, David Warren 204
Salorno 65
Saluzzo 89
Salzburg, city 89
Salzburg, region 6, 17, 28, 37, 50, 67, 139, 143–45, 149, 230–32, 234–35
Sardinia, kingdom. *See* Savoy-Piedmont.
Savoie, department 28, 37, 59, 67, 161, 168, 230–32, 234
Savoy, duchy illustration 7, 16, 33, 36, 62, 69, 147, 149, 156, 161, 163–71, 172–73, 179, 185, 189–90, 199, 206, 215, 223
Savoy, house of 16, 148, 163, 170
Savoy-Piedmont 16, 20, 59, 103, 149, 164, 177, 178
Scheuchzer, Johann Jakob 59
Scheuermeier, Paul 72
Schiera, Pierangelo 214
Schiller, Friedrich 215
Schwaz 86, 89, 92, 94–95, 235
Schwyz 28, 37, 67, 230–32, 234
serfdom 153
settlement/s 49, 53, 56, 64, 75, 77, 85, 91, 94, 97, 109, 115, 145, 146, 156, 162–63, 169, 177, 185, 189, 203, 204, 206, 208, 209, 211–12, 224
Seven Communes 85
sex, masculine/feminine 125, 154, 162–63, 168–69, 171, 174–75, 177–79, 184–85, 189, 204–5, 212
sheep 55, 61–63
shepherds. *See* Hirtenland, Alpine pasture workers, cow-men.

Siddle, David J.   169
Signot, Jacques   illustration 3
Simler, Josias   114
Simmel, Georg   2
Sisteron   89, 238
Slovenia   5, 9, 35, 38, 62, 65, 149, 162, 163, 217
Smith, Adam   166
social orders. *See* estates, provincial.
Sommeregg   183
Sondrio   107, 237
Sondrio, province   17, 28, 36, 37, 40, 67, 161, 230–32, 234
specialization   39, 55, 69–70, 102, 126
St. Gothard   93, 122, 206
St. Gall   28, 37, 39, 67, 89, 99, 230–32, 234
St. Veit   95, 145, 236
stabling   50, 51–53, 56, 62–64, 71, 73, 76, 211
Staffler, Johann Jakob   56, 57, 64
Stanzach   119
state, citizens   153
state, formation   15, 127, 164, 170, 180–82, 190, 200–201, 206, 213, 216, 223–24
state, national   15, 19–20, 107–8, 120, 126–27, 149, 200–202
statistics, method   9, 11, 24–26, 29–34, 39, 43, 85, 91, 103, 104, 108, 138, 139, 143, 144
steepness   55, 57, 64, 70, 74–76, 128, 130
Steyr   89
Stolz, Otto   152
Styria (Steiermark)   17, 28, 37, 40, 41, 58, 65, 67, 75, 105, 107, 123, 143–45, 149, 152, 155, 181, 203, 230–32, 234, 236
subjects   147, 150, 151, 167, 168, 171, 181, 185, 188, 189, 196, 202
Susa   89, 92, 93, 99, 106, 237
Switzerland   5, 6, 19, 25, 33, 35, 38, 61, 66, 69, 121, 137, 139, 150, 153, 172, 179, 206, 213, 215, 217. *See also* Helvetic Confederation

Tarasp   177, 178, 187
Tarentaise   171
taxes   93, 147, 153, 164, 166, 167, 172, 181, 184, 185, 188, 190, 198
technology   70–77, 94, 97, 210
terracing   64, 68, 74, 77, 130
territorialization   15, 16, 175, 190, 199, 208, 212
Ticino   28, 36, 37, 59, 61, 67, 71, 119, 230–32, 234
Tilly, Charles   15, 200–201
timber floating   illustration 10, 120
tithes   65, 150, 167, 174, 179
Törbel   128–29
tourism   3, 13, 108, 110
trade. *See* market.
trade, dependent on density   119–23
trade, livestock   70, 121–23
trade, timber   illustration 10, 119–21
traffic   3, 72–74, 83, 92–94, 96, 119, 121
traffic, river   illustration 10, 99, 119–20
traffic, vehicular   72–74
transhumance   51, 55, 63
transport with back-carriers   72–73
transport, system of transport   69, 70–75, 77, 84, 93–94, 97, 99, 106,

109, 117–20, 122–24, 131, 206, 212, 226
transporting soil uphill   74
Trentino, province of Trento   20, 28, 37, 40, 50, 65, 67, 69, 72, 138, 139, 143, 144, 152, 153, 212, 230–32, 234–35
Trent   6, 7, 17, 89, 92, 99, 237
Turin   16, 106, 148, 164–65, 167–68, 171, 199
Tyrol   6, 17, 20, 34, 40, 50, 56, 64, 65, 67, 69, 72, 92, 119, 138, 143, 145, 149, 152, 155, 157, 176, 208–9, 212
Tyrol, Latin. *See* Trentino.
Tyrol, northern   20, 28, 37, 40, 50, 65, 67, 69, 145, 153, 230–32, 234–35
Tyrol, South. *See* Alto Adige/Südtirol.

Ubaye   20
Udine, province   42
Unterwalden   28, 37, 67, 230–32, 234
urbanization. *See* cities, growth.
Uri   28, 37, 67, 230–32, 234

Val Camonica   72
Val di Non   212
Val Pusteria   56
Val Sesia   72
Val Varàita   162
Val Venosta   56
Valais   17, 28, 37, 40, 57, 58, 67, 128, 150–51, 162, 163, 230–32, 234
Valence   147
Val d'Aosta   28, 36, 37, 67, 74, 92, 144, 230–32, 234
Valsassina   125
Valtellina   36, 59, 68–69, 172

vegetative period (growing season)   55, 59–60, 98, 118, 128, 131
Veneto   16, 50, 63, 101, 125, 149
Venice   16, 62, 86, 119, 121, 124–25, 196
Vercors   58
Verona   89, 99
Veyret, Germaine and Paul   7
Viazzo, Pier Paolo   13–14, 24, 66, 123
Victor Amadeus II of Savoy   164, 166, 171
Vienna   7, 16, 17, 120, 149, 151, 180, 196, 198
Vienne   126
viticulture   49, 52, 55, 58, 68–69, 107, 210, 211
Vorarlberg   6, 28, 37, 50, 56, 65, 67, 69, 75, 139, 143, 144, 152–53, 230–32, 234–35

Walachia   121
war, against the Turks   181
war, Italian wars   16, 174, 202, 206
war, Thirty Years' War   36
wills   154, 168–71, 190
Wilten   105, 237
Wipptal   56
women. *See* sex.
work, seasonal   128, 130
workload   48, 58, 62, 64, 68, 69–70, 77, 81, 116, 186
writing   170, 176, 177, 181
Wunder, Heide   136
Württemberg   204
Zurich   16, 59

www.ingramcontent.com/pod-product-compliance
Lightning Source LLC
Chambersburg PA
CBHW050136240426
43673CB00043B/1687